Trail Guide to World Geography

A Teacher's Manual

With Daily Geography Drills

Cindy Wiggers

Revised 2006
Published by Geography Matters, Inc

Cover design by Alex Wiggers
Photography by Kim Whitmore-Weber

ISBN: 978-1-931397-15-5
Library of Congress Control Number: 2002092174

Printed in the United States of America

Geography Matters, Inc
800.426.4650
www.geomatters.com

Trail Guide to World Geography

By Cindy Wiggers

Table of Contents

Geography Through Literature

Appendix

Acknowledgments

Special thanks to Sylvia Hemme of CAPERS for permission to use her *Around the World in Eighty Days A Unit Study Guide for Homeschoolers* as the basis for the literature unit included in this book.

Thanks to Janet Farmer for her valuable input and editing of the original manuscript.

To Mary Jo Tate, who meticulously edited the 2005 reprint, thanks. Your valuable input and professionalism brought much needed clarity to our little corner of the world.

Thanks to my daughter, Libby, for sharing her skills with layout and design. We could have never kept on schedule without you.

To my dear husband, Josh, who designed a special map for this book and assisted in ways too numerous to list, I offer thanks from the bottom of my heart. You are a gift from God.

To Josh and our son, Alex, who willingly helped me with the daunting task of reviewing every question in the 2006 revision to check for accuracy, edit for world changes, and to adapt wording to match the changes that occurred when the recommended atlases were reprinted. You are both awesome!

Acknowledgments

Special thanks to [illegible]

Thanks to [illegible]

In memory of [illegible]

Trail Guide to World Geography

Introduction

Wouldn't it be exciting to turn your geography studies into a guided trip around the world? How wonderful just to sit back and relax as the guide shows the way and takes your students through a learning experience they will never forget! I understand how difficult geography is to teach when you know little about it yourselves, have major time crunches, or when you are trying to teach several grade levels at once, so I've written this manual to be just the solution you seek.

The *Trail Guide to World Geography* will take your students on a journey to each continent. As they blaze the trail, there are many interesting topics to explore. Students will soon be able to recognize important characteristics of each continent, read and create maps, identify key geographical terms and more, while they work through a variety of quick daily lessons or dig into in-depth geography studies.

First, it's important to understand a bit about what the study of geography encompasses.

> Geography is a knowledge of place names, location of cultural and physical features, distribution and patterns of languages, religions, economic activities, population and political systems. Physical regions and physical phenomena, such as tectonic activity, landform, climate, bodies of water, soils, and fauna. The changes in places and areas through time, including how people have modified the environment. Cartographers' tools, such as maps, instruments, graphs, and statistics, are also a part of geography.
>
> —National Geographic Society

Throughout the school year, you'll address all of these aspects in one way or another. This simple-to-use plan includes five to ten-minute geography drills, mapping assignments, and appropriate hands-on activities which use resources that are readily available. You choose how much or little you want to cover.

Each of these three sections is explained in detail in the following pages:

- Geography Trails
- Points of Interest
- Geography Through Literature (a unit study based upon Jules Verne's *Around the World in 80 Days*)

May you have a fun-filled year of geography! Oh, and pick this book up again next year. Your students can do the Geography Trails (the atlas drills) from a different level to keep sharp on their geography terms, place location, and atlas usage.

Cindy Wiggers

1. Geography Trails

This manual is divided into three sections. The first is the daily atlas drills, called Geography Trails. These drills can be used alone to introduce and reinforce geography terms and place-name recognition, or they can be used as a component of a full geography course by using the activities and assignments from the other two sections. Select from three levels of questions identified by the following trail markers:

Primary: 2nd-4th grades Intermediate: 5th-7th grades Secondary: 8th-high school

Here's how it works: Students answer two questions a day, four days a week, providing consistent geography moments throughout the school year. The objective is to build students' knowledge and awareness of physical and political world geography. Your students will receive daily practice in recognizing geography terms and using maps to locate landforms, bodies of water, countries, and capital cities. The short daily repetition of skills helps develop long-term recall of places and facts. Faithfully performing these daily exercises lays a solid foundation for geography studies.

Geography Trails Objectives

Primary	To introduce maps and their usefulness To teach students how to use maps and read a map legend To introduce students to outline maps To guide students in the use of an atlas To introduce a variety of geographical terms To recognize the difference between continents, countries, and cities To lay a foundation for students to be able to use maps and atlases independently by the end of third grade or beginning of fourth
Intermediate	To develop independent map-reading skills and atlas usage To be able to locate any place in the world by using an atlas To become familiar with physical features of the world To increase understanding of geographical terms To identify continents, countries, and cities on a map To introduce critical thinking skills using maps
Secondary	To develop proficiency at reading maps To become skilled in using atlases To know the location of important places in the world To know key world physical features To understand geographical terms To develop critical thinking skills using maps and an almanac

Scheduling Geography Trails

This course allows for flexibility when planning your school schedule. You may want to start your history or social studies each day with these five-minute geography drills. Another option is to assign them at the beginning of the day, while students are gathering their thoughts, to focus on schoolwork again. Ideally, students should perform this task at the same time each day. To improve retention of facts and develop geography awareness, it's more effective if students work five minutes a day for four days than twenty minutes in one day. If at all possible keep to a four-day routine.

2. Points of Interest

The Geography Trails provides the quick hike on the trail of least resistance. This is perfect for some treks, but for others a journey throughout the world wouldn't be complete without slowing down to view special Points of Interest or blaze a trail to enjoy the scenery along the way. That's the idea of this section, which includes these two parts, explained in detail starting on page 13:

- **Mapping** places and features on outline maps
- **Trail Blazing** additional research and hands-on activities

Points of Interest Objectives

Primary

To begin to learn countries and capitals
To improve memory retention of geography learned
To introduce outline map usage
To instill a joy of learning through a variety of procedures, art projects, and other hands-on activities

Intermediate

To improve memory retention of geography learned
To develop skills with using outline maps
To learn countries and their capitals and where they are located
To introduce a fuller understanding of geography through the use of thematic maps, research on a variety of topics, and exposure to world cultures
To instill a joy of learning through a variety of procedures, art projects, and other hands-on activities

Secondary

All objectives in the intermediate level PLUS,
To increase critical thinking skills by analyzing thematic maps, researching a variety of topics, and gaining exposure to world cultures

Scheduling Points of Interest
This teaching guide is flexible to meet your objectives and scheduling needs. Depending upon how much work you assign, students could work about half an hour daily on a Points of Interest section for four to five days a week. Or you may decide to do the five-minute Geography Trails drills Monday through Thursday and select from the Points of Interest assignments for Friday. Consider reserving a full class period for art projects or any other assignment that would benefit from the extended time. Set aside a few minutes each Friday for students to show their work and explain what they have learned through the week

3. Geography Through Literature
The final nine weeks of the school year are reserved for reading the classic novel *Around the World in Eighty Days*. By this time you have studied all the continents. Now students will travel the world with Phileas Fogg and Passepartout, mapping their journey and choosing from a variety of additional assignments and activities including research, crafts, spelling, and vocabulary. More detailed instructions begin on page 85.

Geography Trails Instructions

Geography Trails are designed to be used as daily five to ten-minute atlas drills for students capable of reading an atlas. Make sure you provide all students, regardless of level, instructions on using an atlas and reading maps. Use the first two or three weeks to hike the "trails" together to either teach map skills or as a review. The Geography Trails should become a five-minute daily exercise when students have had enough experience to confidently find their way around the atlas without your help.

The daily drills focus on a specific region of the world each week for twenty-seven weeks of the school year. Choose from three different trails according to the appropriate academic level of your students. Starting with an overview of the world and moving around the globe continent by continent, students simply follow the trail mark each day that is associated with their level. These three levels address map reading skills and comprehension at graduated levels from basic to more difficult. Although I have suggested grade levels for each trail, it is best to select the trail according to your students' reading comprehension, map usage proficiency, and reasoning abilities, not according to his or her actual school grade.

Each of the three trails is marked by designated animal tracks.

Represents questions at the primary level. Teachers of students not yet reading can use this level to introduce daily geography moments into the school schedule. Find places on a map, and answer questions aloud together. Students competent with reading can generally answer them without help, if sufficient instruction in using maps and atlases has been provided. Most questions for this level can be answered by using a primary atlas, in particular, the *Beginner World Atlas* by Rand McNally.

Represents questions intermediate students can handle alone. Most questions for this level can be answered by using an elementary atlas, in particular, the *Intermediate World Atlas* by Rand McNally.

Represents questions directed to secondary students. Most questions for this level can be answered by using an advanced atlas, in particular, the *Answer Atlas* by Rand McNally. To further challenge advanced high school students, you may want to assign questions from both and .

Note: Secondary students should be developing solid skills with research; therefore answers to a handful of Geography Trails questions may require use of an almanac, the Internet, or other resources. These few questions are scattered throughout the course and may take the student more than the average five to ten minutes to complete the daily drill. The *Answer Atlas*, with its detailed facts sections and charts in the front of the book, will contain answers to many of the Geography Trails questions.

Resources Used with Geography Trails
All that is needed to use this section of the *Trail Guide* is a student atlas appropriate to the student's ability and perhaps a current almanac (for the secondary level). For ordering information see page 126.

Recommended atlases by level are:

Primary	*Beginner World Atlas*
Intermediate	*Intermediate World Atlas*
Secondary	*Answer Atlas*

Geography Terms
For questions regarding geography terms use any of the following:

- the glossary in the student atlas
- a geographical terms chart
- geography flash cards in *The Ultimate Geography and Timeline Guide*
- Appendix A in The *Ultimate Geography and Timeline Guide*
- a dictionary

It's as simple as that. Two questions a day for four days each week, and your students develop a general consciousness of the world around them. If you are teaching more than one level at a time, all students can learn about the same area of the world together. Answers for all levels are located in the back of the book.

Geography Trails can be used at least three separate years with one student. Questions are not generally repeated but are more in-depth at each increasing level. If you use this book solely for the **Geography Trails** daily drills, you've gotten your money's worth already. However, there is much more here to assist you in teaching geography, so expand your horizons and open the way for the additional activities provided in the **Points of Interest** section.

Points of Interest Instructions

The daily drills are great for providing regular geography moments and to lay a foundation for geography knowledge, but the real nitty gritty of your geography studies will be found in the **Points of Interest** section of the *Trail Guide*. You have a wide variety of assignment choices here, so match your student's learning style and interests and you're sure to create a fun learning environment. This section is divided into two parts: **Mapping and Trail Blazing**.

Mapping

Assignments in the mapping section provide the opportunity for students to make their own maps from basic outline maps. The outline map title needed is given above the assignment list. When (OMB) follows a map title, it is a reference to the page number in *Uncle Josh's Outline Map Book* or Uncle Josh's Outline Map Collection CD-ROM. If you use Uncle Josh' s maps on CD-ROM with a copyright date before 2008, you may need to add 3 to the page number to find the same map.

One outline map is included on page 108 in the back of this book to use while studying Asia. It covers the area of old U.S.S.R. Republics that broke away to form several independent countries. The map scale of this area was too small on most outline maps, so one is provided in the appendix for your convenience. Feel free to make copies as needed. (Uncle Josh's Outline Maps on CD-ROM does include the map of this region. Uncle Josh named it "Stan and His Sisters"—Uzbekistan, Kurgyzstan, Turkmenistan....)

Most students really enjoy filling in their outline maps. It is fascinating to see the personality and individuality of each student reflected in the quality of their maps. However, let me share a word of caution with you: don't expect students who've never used outline maps to jump into these projects without assistance and encouragement. You may need to work with them the first few times, as a blank map can be a bit intimidating. Show them the similarity in shapes to the maps in their atlas to help them recognize landmasses and bodies of water. Once they get used to performing this kind of mapping exercise most students will take to it with pleasure and a level of pride in their work.

Instruct students to use their best penmanship and to be consistent. For continuity in map labeling, use uppercase and lowercase letters for city names and all CAPS when printing country or capital names. Students may choose to use a dot within a circle for capital cities. (Sometimes placing a star on the map for capitals just takes up too much space.) Whichever students choose, they just need to be consistent. Use colored pencils (perhaps erasable) for shading and fine-tipped markers, if possible, for labeling place names. Use blue for bodies of water.

If you set up a color coding system, you can place it on an index card for reference. For example: blue for bodies of water, green for plains, purple for mountains, brown triangle for mountain peaks, all CAPS for capitals, etc.

Using Atlases

Do you need some instructions on how to use atlases? The front (or back) of any good atlas usually explains its use. Show students the meanings of the map symbols used by the publisher. Not all publishers use the same symbols and colors to represent the same features. However, many are the same. Triangles generally represent mountain peaks, curvy blue lines represent rivers, and so on. Some publishers use a star to represent capitals, others a dot inside a circle, and yet another may underline the name of capitals. Just be sure to look at the legend explanation given in the instructions. Help the students become familiar with the legend so they can put a nice legend on their own maps as well.

Mapping Option

You may choose to have students do the mapping assignments in *The Ultimate Geography and Timeline Guide* instead of those included in the *Trail Guide*. Again, references to these mapping activities, or "Map-It" assignments and the corresponding page numbers, are included for your convenience and look like this:

OR do North America "Map-It" (TUG 160)

OR do North America Outline Activity (TUG 205)

Map work is best done at one sitting, so assign the **Mapping** for one day, and use the **Trail Blazing** assignments for the remainder of the week.

Trail Blazing

As you lead your students on this journey around the world, you can blaze a trail with some additional mapping and more hands-on activities. You have the opportunity to select projects that meet the learning style of your students or to vary the assignments to span the spectrum of learning styles. There is a wide range of assignment choices—more than anyone could be expected to complete in any given school year. They include:

- Research topics
- Projects and activities
- Geography Notebook
- Illustrated Geography Dictionary
- *Geography Through Art*

Select from these projects each week. Please do NOT expect them to do all of the assignments. Unused activities can be reserved for future school years or applied to social studies or science classes.

Find what type of work interests your students. You may want to allow them to choose assignments that pique their curiosity. Let the different studies serve as a catalyst to develop and enrich a love for learning. Geography is so wide a subject that there should be no shortage of opportunities to light a spark of interest in nearly every student.

Research, Project, and Activity Options

Some of the **Trail Blazing** assignments are listed below with more detailed instructions than space allowed in the weekly text. Refer back to these pages when needed. You may want to place a paper-clip on this page as a bookmark to easily return to this section of instructions when needed.

3-D Maps

Students make salt dough maps or 3-D maps of the continent landmasses. *Geography Through Art* (GTA) has several different recipes for clay or dough you can make yourself. Here is a simple salt dough recipe:

2 parts flour
1 part salt
1 part water

Add more water if it's crumbly. Attach a copy of the outline map onto a piece of cardboard. Place dough on map and spread to edge of land border. Use a physical relief map as a reference to form the landforms and shape of the continent. Let dry overnight before painting, or add food dye to dough before shaping.

Flash Cards

Using index cards, students make their own flash cards of countries and capitals. On one side draw the shape of the country and place a star (or a circle with a dot in it) at the location of the capital. On the back write the name of the country and its capital.

Optional: put the name of the capital on the drawing next to the star and the name of the country on the back. While the students are learning the name of the country they are already "seeing" the name and location of the capital. Either side of the card can be used to drill and either side can be used as the answer.

Foods

Students are encouraged to prepare and eat international foods. Specific countries are listed, but do not be limited to any list. The idea is to expose students to different cultures through their foods.

Principal Crops or Natural Resources

Students are instructed to learn the principal crops or natural resources of an area. Be sure they note the climate, soil, and location whenever possible. See if they can recognize the effect on the economy. Often this information is on the thematic maps in your atlas. Other sources: almanacs, encyclopedias, library books, the Internet, and more.

Principal Cities and Population

For additional critical thinking, help students recognize the association of population densities with natural resources, transportation, or availability of water. To do so have them study the population density, climate, and natural resources thematic maps in the atlas.

Landmarks, Buildings, and Engineering Achievements
Students will research more about these topics and include pictures in their notebooks. Study construction techniques and how the land, soil, and climate were taken into consideration. Listed are just a sampling of famous buildings of significance. You may wish to add your own. Just let assignments of this type serve as a reminder to dig deeper and to help students realize that even something like the study of architecture or construction techniques is influenced by and contributes to the geography of our world.

Jigsaw Map Puzzle
To make a jigsaw map, have students color an outline map of the continent, using different colors for each country. Cover with contact paper. Cut into different shapes or cut along country borders. Once the puzzle is made, let students put it back together. They can time themselves to gauge their own progress. Store puzzle pieces in zipper-sealed plastic bags or in sheet protectors placed in the Geography Notebook.

Another option is to make a large puzzle with card stock or make a large map of the continent by enlarging in a copier. Tape all the pieces together, color, attach to poster board, cover with contact paper, and cut out. (For more map-making ideas see *The Ultimate Geography and Timeline Guide* pages 61–64).

Geography Concentration
Cut index cards in half. Write the name of a country on one card and the name of its capital on another card. Continue for all countries on the continent of focus. Play "Memory" or "Concentration" to practice learning countries and their capitals. Shuffle cards and place in rows blank side up. Players turn up two cards at a time, trying to match country and capital. When a match is made, the player keeps the cards and goes again. When all cards have been matched, the player with the most cards wins.

Animal Studies
Because animal life depends upon the physical characteristics of a place, students should learn about the animals of each continent. They can develop an animal notebook or establish an animal section in the Geography Notebook. Student will be reminded to include new animals as they learn about each continent. They should begin to notice, for example, the variety of animal life in Africa that is not indigenous elsewhere in the world and relate this to how climate, soil, and water affect biological life. Include pictures and drawings when possible.

Thematic Maps
Students will make their own set of thematic maps of each continent. Have them choose from the following map themes: physical elevations, land usage (environments), population, climate, natural hazards, or whatever thematic maps are provided in their student atlas.

1. Make several copies of the continent outline map.
2. Duplicate the theme from the atlas by drawing and shading the different regions with appropriate colors.
3. Place a legend box in an unused corner of each map to show what each color represents.

Crossword Puzzles
Have students make their own crossword puzzles using the country as a clue and its capital as the answer. They can also use other information such as landmarks and famous places. Make copies of their crossword puzzles and share with others. Put one copy of the puzzle and its answer key in the Geography Notebook. Some students like to draw pictures on the page expressing their individuality.

Country Statistics
Make a chart of continent facts. Place country names down the left side of the chart. Add columns across the top and choose labels from these topics: capital, area, currency, language, principal religion, and natural resources. This should look like a spread sheet. See how many cells in the chart students can fill in from available resources, the Internet, or an almanac. (These basic statistics can be found in any almanac.)

Newspaper Clippings
Students should watch the paper and magazines for current events in the continent of focus. Clip the articles for inclusion in the Geography Notebook. Use the articles as an opportunity to discuss world events and their effect on others.

Making a Geography Notebook

Each student can create his own unique geography notebook from the Points of Interest projects he completes. This will become a permanent record of the geography study. Use a three-ring binder with tabs for each continent. Place maps, pictures, written reports, flags, and other projects in the appropriate sections. Create a separate section for the Illustrated Geography Dictionary. Place flash cards in a plastic sheet protector or in any variety of three-ring pocket dividers, available from your local office supply store. Organize the notebook in whatever way suits you or your student. For more detailed information on making a student notebook, see instructions in *The Ultimate Geography and Timeline Guide* pages 14–19.

Some additional routine projects students may faithfully add to the notebook include: their own thematic maps, charts of country statistics, additional "Map-It" projects from *The Ultimate Guide,* articles clipped from the newspaper or magazines, completed "Countries of the World" fact sheets from the *Geography Through Art* book, research, writing, and more.

An optional student notebook designed specifically for each level of this curriculum is available in your choice of eBook or CD-ROM. This printable book contains all of the outline maps you'll need for the mapping assignments, plus weekly pages with the Geography Trails questions and a place to write the answers. In addition, there are numerous formatted pages or templates for assignments, drawings, crosswords, and more. The *Trail Guide to World Geography* Student Notebook files save time and planning and make creating a geography notebook a cinch! For more information and sample pages, go to www.geomatters.com and select eBooks.

Illustrated Geography Dictionary

Students can create their own colorful "Illustrated Geography Dictionary." They will add new words and terms throughout the first half of the year using the Illustrated Geography Dictionary pages included in the appendix.

Make copies of pages 112-116. Instruct students to draw images of the various geographic features assigned each week in the appropriate box on the sheet. If they need help to draw the features, consider providing an illustrated chart of geographical terms as a handy reference. Write a short definition of the feature under the picture. There's a completed sample on page 111 and also a reproducible blank page for any additional terms. For basic definitions use a geography terms chart, the glossary in the student's atlas, or refer to the geography flash cards and definitions in Appendix A of *The Ultimate Geography and Timeline Guide*.

Geography Through Art

Learning about the art of places around the world is a wonderful way to introduce world cultures and this connection helps deepen understanding of the geography. *Geography Through Art*, by Sharon Jeffus and Jamie Aramini, is loaded with interesting and creative, do-able projects that integrate geography and culture through art. Most projects include background information related to the history and people where the project originated. At the bottom of Points of Interest pages, you'll find the titles of projects from the *Geography Through Art* book that are appropriate to the region of study for that week.

If you like the idea of incorporating geography with art, or if you have students who respond well to art, you'll want to select assignments from this section regularly. Plan ahead for supplies and allow ample time for students to do the activities. Display and show off their handiwork at every opportunity. Photograph finished projects for inclusion in the Geography Notebook. It's amazing how much more kids remember when they get to do something hands-on associated with their studies. Too much time in front of textbooks is boring, but add an appropriate project and watch that delight for learning come alive in the sparkle in their eyes!

Three Key Resources

Since this is not a textbook, assignments in the Points of Interest section require the use of a variety of additional resources. Students will use research materials such as an almanac, encyclopedia, library books, Internet, and videos. Three specific resources that you should consider having handy if you want the full benefits of this Trail Guide are listed below. These additional resources are:

The Ultimate Geography and Timeline Guide
Geography Through Art
Uncle Josh's Outline Map Book or CD-ROM (or a good set of outline maps)

1. The Ultimate Geography and Timeline Guide

Have you ever had a book that was the perfect resource for a project, but didn't realize it until after the project was completed? Maybe you wasted valuable time and energy researching information that was right there on your desk or bookshelf all along. Either you didn't realize it or forgot all about this information being in that book. If you own a copy of *The Ultimate Geography and Timeline Guide* (*The Ultimate Guide*) you may be using it regularly. (I hope so!) Or, you could be like those who don't realize just what a treasure trove of information is sitting there on the bookshelf waiting to be used.

The *Ultimate Geography and Timeline Guide* has loads of information that can be used both as a reference for the student and also for instructions to the teacher. It is especially helpful to anyone needing detailed information on geography and instruction on how to teach geography. Rather than repeating or rewriting that same information in this *Trail Guide*, I have referred you to it whenever information from the *Ultimate Guide* would be an additional benefit to the lesson at hand. It is referenced with page numbers for your convenience. The reference looks like this (TUG 23). Also, as mentioned before, the flash cards and definitions in The *Ultimate Guide* are great tools for learning geography terms and for answering *Geography Trails* questions associated with geography terms.

Some Points of Interest projects are taken directly from *The Ultimate Guide*. However, there are plenty of other options to choose if you don't own a copy of the book. Please understand that you do not need to go out and buy *The Ultimate Guide* in order to use this *Trail Guide to World Geography*. It's an added bonus if you have it.

2. Geography Through Art

If the idea of learning the culture of the world through doing art projects influenced your selection of this book, you'll want to obtain a copy of *Geography Through Art*. All art projects are selected from this book to enhance your study of world geography. It's also used as an additional reference book in the Trail Blazing section. If you see (GTA), it means there is a study on that topic in *Geography Through Art*.

3. Uncle Josh's Outline Map Book (or CD-ROM)

You need a source for reproducible outline maps for your students to write on. If you don't have maps yet, please consider *Uncle Josh's Outline Map Book* or Uncle Josh's Outline Map Collection

CD-ROM.
My husband and daughter designed the outline maps to meet our guidelines for quality and have these features not usually provided by other map sources:

1. The rivers and bodies of water are lightly shaded.
2. All places are shown in context with surrounding boundaries.
3. Grid lines for longitude and latitude are included on many maps.

These features are especially helpful to students as reference points when finding where to put assigned information and often eliminate frustration. With outline maps that include rivers, students simply select the appropriate river, trace it, label it, and move on. Students who tend to be perfectionists are relieved of the internal pressure to draw rivers exactly as they see in their atlas. These maps are available in book form for photo copying or on CD-ROM for easy printing from your home computer. Again, it is not necessary to have this exact book, but you do need a source of good outline maps.

Reference codes
For your convenience when an assignment topic is covered in *The Ultimate Guide* (TUG), *Uncle Josh's Outline Map Book* (OMB), or *Geography Through Art* (GTA), the book code and page number may be given in parentheses. For example, a Trail Blazing assignment may read:

> "Read about North America (TUG 160–168, GTA, student atlas), and find travel videos from the library of places located in North America."

Your students can read about North America from their choice of any variety of resources. *The Ultimate Guide* covers North America on pages 160–168, there is a summary of North America (and each continent) in *Geography Through Art*, and most student atlases provide information on each continent.

In most cases it's not absolutely necessary to have these books in order to assign the projects, because the information can also be found in other research materials. It's just more convenient to have them handy if your budget allows.

The assignment ideas in the Points of Interest section only scratch the surface of what can be done while studying the continents and countries of the world. Add your own ideas, adapt ones given here, and enjoy learning right along with your students!

Additional Recommended Resources

This section describes some of the additional recommended resources you will find helpful in teaching from this guide. Your geography lessons will go much smoother if you have the basic resources available. Unless otherwise indicated, resources listed below can be used by all level of students.

Atlases

It is important to consider font size, complexity of maps, ease of use, and information provided when selecting the atlas that best fits each student's need. If you use these atlases your students should be able to find most answers to the Geography Trails atlas drills in about five to ten minutes. These were the atlases used to write the questions.

Beginner World Atlas
A very basic primary level atlas. Excellent uncluttered maps and photographs of places around the world.

Intermediate World Atlas
Clear, concise maps recommended for intermediate students. Excellent source of thematic maps, geographical terms, historical timeline for each continent, photographs, and special informational sections. Upper level students using the *Intermediate World Atlas* will also need an almanac or other book of facts to answer some of the Geography Trails questions.

Answer Atlas
Uses highly acclaimed map art from *Goode's World Atlas*. Recommended for secondary students. Includes thematic maps and continent summaries. Multi-layered maps may seem too cluttered, but added world information sections make this a worthwhile atlas. This book's question-and-answer section and informational charts are a handy resource to answer questions on the Geography Trails section and for research.

U.S./World Laminated Map

Most effective when guiding second and third graders through the mapping assignments, but useful at all ages. Reusable large-scale laminated map designed for write-on and wipe-off flexibility. The U.S. is printed on one side and the world is on the reverse. (23″ x 34″)

Geographical Terms Chart

The color illustrated physical features laminated chart has over 150 terms defined on the back. Great for all levels as a visual aid and for geography vocabulary. Students will use this as a reference to draw physical features in the Illustrated Geography Dictionary section of their notebook. It is also used for answering terms related *Geography Trails* questions.

Usborne Encyclopedia of World Geography (with complete world atlas)
This internet-linked book is a valuable reference for a great number of research assignments throughout the Trail Blazing sections including planet earth, earthquakes, volcanoes, rivers and oceans, weather, peoples of the world, and more. Each page includes Internet links to Usborne's website pages that expand studies further. In addition, it has information on reading maps with a complete world atlas in the back of the book.

Eat Your Way Around the World
To provide a culture focus in any geography study, students love eating international food. This book contains meals from around the world, catagorized by continent and is referred to in the Trail Blazing section. More than a cookbook it includes of meals (main dish, side dish and dessert) from 30 countries. In addition to recipes with ingredients that are readily available, students learn international etiquette and culture tips.

Almanac
An almanac is often the best resource to use with some fact-finding questions and research at the more advanced level. Although almanacs are updated annually, most data is still pertinent with almanacs that are a couple of years old. If students have not had sufficient exposure to this valuable resource, you may want to set aside a day or two for providing additional instructions in using an almanac (see TUG 21–22, 240). For additional practice make up some of your own questions with answers that can be found in the almanac. (The *Trail Guide to U.S. Geography* uses an almanac as the main resource for answering Geography Trails questions at the secondary level.)

Supplies
It's helpful to have the following additional items handy before beginning this program.

Dictionary
Three-ring binder with dividers
Colored pencils or fine-tipped markers for use on paper maps
Vis-a-vis erasable pens for use on optional laminated map
Large-scale continent outline maps (optional)

Lesson Plan Suggestions

These lesson plans are offered as a suggestion to get you started. Use them as a guide and remember that this curriculum is meant to be flexible to meet your needs.

Spend 5–10 minutes each day for four days on the **Geography Trails** daily drills. On day five assign 20–30 minutes (more if your schedule allows) on your choice of activities. Select from **Mapping** activities or **Trail Blazing** projects. Choose projects that will enhance student enjoyment of learning geography. Some projects are best done as a class or in a group; others require assistance. Adapt any assignment choice to meet your students' abilities. Plan ahead for projects that require additional supplies.

Scheduling

Depending upon how much work you assign, students can work about half an hour daily for four days. Reserve a full class period for art projects, or any other assignment that would benefit from the extended time period. Set aside a few minutes Friday for students to show their work and explain what they have learned that week.

A sample general weekly schedule follows. Use this as a guide. Adapt it to meet your own objectives and time frames.

Monday
5–10 minutes on Geography Trails
20–30 minutes Mapping

Tuesday, Wednesday, and Thursday
5–10 minutes on Geography Trails
20–30 minutes Trail Blazing

Friday
Trail Blazing
Show projects and tell what has been learned
Final copy of any written work due
Complete any unfinished assignments

Answers

Answers to Geography Trails questions are located in the back of the book. Every effort was made to ensure accuracy at the time of publication. Please feel free to submit any corrections to the folks at Geography Matters.

Combining This Course with History

Making geography's connection to history is natural and also deepens your student's understanding of both subjects. This book is a great additional resouce to use alongside any history study. If you want to focus on the geography of the region you are studying in history, simply ignore the week numbers and study each continent in your own way at your own pace.

Selecting Levels
Choose the level based more upon how much exposure your students have had to geography and map usage than to their academic grade level. Use the lesson plans as a guideline. Be flexible and sensitive to the needs of the student. Learning should be a fun experience for children. This will contribute to developing individuals who can think for themselves and who are lifelong learners.

Primary: Second–Third Grades
(or until students are reading well)

The primary level is a teacher-led course with the objective of developing the student's ability to work alone by the time he is ready for the intermediate level. Three weeks of specific instructions for the primary level follow. That should be enough to get you going so you will understand how to proceed for the rest of the year. (Note: The primary level questions can also be used by any level student for simple daily geography drills and atlas usage. Please do not limit yourself by these levels.)

On Monday through Thursday, use the daily questions as a framework to introduce geography terms and place recognition. Always teach the meaning of new words. Use visual aids where possible. (An illustrated geography terms chart is very helpful.) Significant geography terms are listed on the Points of Interest page for the first 12 weeks.

Read the questions and find the answers together. Show students how to use the atlas and read to them anything they have not yet developed the skills to read for themselves. At this level you will do all work together aloud.

Locate and show places on a map. Using a large laminated outline map of the world and erasable overhead projector pens, use the questions as a basis for demonstrating information in a visual way. Be creative. Use your own ideas. Here are some examples that coincide with the first three weeks of **Geography Trails** questions:

Week 1
Use the geography terms as vocabulary words. Students who are learning to print can copy the words on a sheet of paper. You can use these words and their basic definitions as an exercise in penmanship practice. Terms used: east, west, globe, continent, latitude, equator, ocean.

Day 1 - Show students a globe. Discuss the word sphere. Point out the large landmasses and their shapes.

Day 2 - Show students the continent shapes on both a globe and a world map. On a laminated outline map of the world, outline the shape of each continent. Point out the oceans.

Day 3 - Name and label the continents. Highlight the continent where you live.

Day 4 - Point to lines of latitude and longitude, and label the equator.

Week 2
Terms used: boundary, longitude, prime meridian, international date line

Day 1 - Show students the oceans on both the globe and map. Name and label the oceans on a world outline map. Point to an ocean on the terms chart if you use one. Point out the United States and label it. Explain the directional terms and put a compass rose on the map. Show which U.S. boundary is east. Help them identify the Atlantic Ocean there.

Day 2 - Point out the lines of longitude. Explain hemispheres and how to identify east and west. Using the world outline map, identify the line representing the zero° starting point or prime meridian (near Greenwich, England).

Day 3 - Show and explain the legend in an atlas and help students to recognize the symbols. Let students answer the questions after they have seen the legend.

Day 4 - Look at the outline map you have been labeling. Can students identify which continent is Australia? Can they identify the oceans surrounding Australia? Review the Northern Hemisphere and Southern Hemisphere.

Week 3
This week you will begin a five-week focus on North America. This continent includes Greenland, Canada, United States, Mexico, Central America, and Caribbean Island nations. Locate and label places on the map as they come up in the questions. Continue learning geography terms: hemisphere, country, island, and sea.

Day 1 - Point out North America. Let students determine what country looks the largest. Label Canada.

Day 2 - Draw a picture of an island from a bird's-eye view. Locate islands on the world-map. Can students identify the oceans surrounding Canada from looking at the world map? Show an island on the terms chart.

Day 3 - Outline Greenland on the world map. It is in North America, but is claimed by Denmark. You may want to point out the location of Denmark. Locate and label Mexico.

Day 4 - Use an atlas. Show the name of the sea off the coast of Alaska (Bering). Point out Asia on the west of the sea. Can students remember the name of this continent? Define sea and show it on the terms chart.

Continue in like manner each day Monday through Thursday, using the questions as a guide to what to teach. Take no more than five to ten minutes each day. Use the student atlas as often as possible to answer questions, continuing to teach the use of the atlas. Even students who don't read yet will gain the concepts to put into practice as reading skills develop.

Primary: Third–Fourth Grades

Students who can read and have been sufficiently taught to use maps and atlases can begin to answer the questions for themselves. Always encourage and praise independence, recognizing the need to guide them only until they are ready to work on their own. Be sure each student has his own atlas.

Galloping the Globe

This is a wonderful resource for teaching geography to the younger set. If you are teaching a wide range of ages and your oldest is fourth grade or under, you may want to start with *Galloping the Globe* and use the *Trail Guide* the following year. *Galloping the Globe* is a kindergarten–fourth grade geography unit study incorporating literature (great annotated book lists for each country studied!), science, Bible, and history. It can easily be used with the *Trail Guide* if you study the continents at the same time, letting the older ones do assignments from the *Trail Guide* while the younger ones color country flags, do mazes, crosswords, and more from *Galloping the Globe*.

Intermediate

If students at this level have had little or no experience in using maps, consider using the first three weeks to review map use, or start with the primary level. Read the front of the student atlas together or use *Discovering Maps* to teach map usage.

Mapping OPTION

Chapter thirteen of *The Ultimate Geography and Timeline Guide* has an excellent geography unit directed at the intermediate level. You may choose to integrate this unit with the Points of Interest projects. If so, follow the trail markings and choose between the mapping projects in either this Trail *Guide to World Geography* or *The Ultimate Geography and Timeline Guide*. Do not require students to do both.

Trail Blazing Hints

Many activities are repeated for each continent, including making flash cards, thematic maps, charting facts, and more. For fascinating ongoing projects, be sure to assign these similar projects for each continent. For example, if you have your student make thematic maps for North America from Week 3, select this same assignment again in Week 9 when it is listed for South America and week 11 for Europe and again for each of the other continents. The student notebook would then have thematic maps for every continent.

Secondary

Lesson plans suggestions for this level are the same as the fifth–seventh grade above. In additon keep the following in mind:

Mapping OPTION
You may choose to integrate the high school level outline activity in chapter thirteen of *The Ultimate Geography and Timeline Guide* with Points of Interest projects. If so, follow the trail markings and choose between the mapping projects in the two books. Do not require students to do both.

Geography Trails
Answers to some *Geography Trails* questions at the most advanced level may not be found from maps in an atlas. This provides an opportunity for students to do a little bit of research outside of reading maps. These kind of questions can be answered by using the front informational section of the *Answer Atlas* or by using an almanac. If you don't want students to labor over the few questions that may not be in the atlas you're using, simply provide the answer from the answer key. They'll still be learning no matter how they got the answer!

Let's Get Started

Okay, you have your instructions, your teaching manual, and your resources, so it's time to get started. Let's quickly re-cap a couple of things:

To use the **Geography Trails** sections (the atlas drills) of the *Trail Guide* you only need the *Trail Guide* itself and a good atlas, geared to the student's ability or level of understanding. Your students will gain valuable skills and insight into the geography of our world when they use this section of the book.

For amore comprehensive geography study, assign the **Points of Interest** activities. You will need the additional resources described on pages 19–22. Many of you already have these on hand, so you're ahead of the game. If not, consider making the investment, or just start collecting these wonderful resources as you can, and work with what you have. Ordering information can be found on page 126.

The goal is for you and your students to become excited about geography! This guide is intended to help by making what can be a challenging subject for many, become routine. When you're ready to study U.S. geography, consider the *Trail Guide to U.S. Geography*.

We've established a Yahoo discussion group for users of the *Trail Guide* series. It's also a great encouragement for anyone teaching geography. For additional teaching ideas, feedback from others, and discussion join us by logging on to:
http://groups.yahoo.com/group/geographytrailguides/

You may be interested in receiving Geography Matters' newsletter in your email box. Bimonthly issue include current geography updates, free map or project downloads, product coupon specials, and additional ideas to help make learning fun. You can sign up from the website at:
www.geomatters.com

Happy Trails to you and your students as you venture around the world with the Geography Matters® *Trail Guide to World Geography*.

Author's note:
I am genuinely interested in your feedback, corrections, and suggestions.
Feel free to contact me at: cindy@geomatters.com.

Notes:

GEOGRAPHY TRAILS

Week 1 - World

Day 1

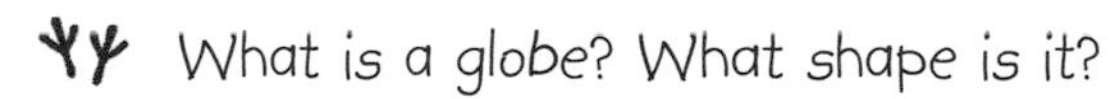

- What is a globe? What shape is it?

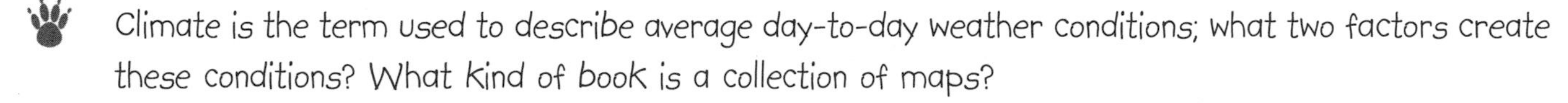

- Climate is the term used to describe average day-to-day weather conditions; what two factors create these conditions? What kind of book is a collection of maps?

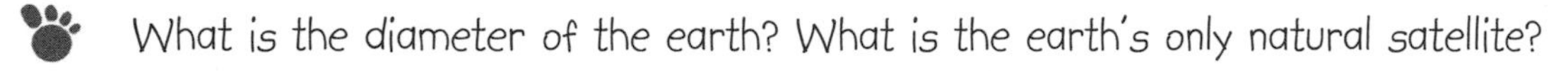

- What is the diameter of the earth? What is the earth's only natural satellite?

Day 2

- What are the earth's largest bodies of land called? What are the earth's largest bodies of water called?

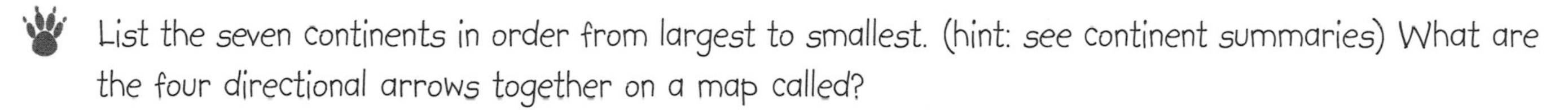

- List the seven continents in order from largest to smallest. (hint: see continent summaries) What are the four directional arrows together on a map called?
- List the seven continents (with their areas) from the largest to the smallest. What is the total land area of the earth?

Day 3

- Name each continent. On what continent do you live?
- What are the lines running east and west across the maps? What continents are located mostly in the Northern Hemisphere?
- What imaginary parallel lines are numbered in degrees north and south of the equator? What continents have some of their landmass in the Southern Hemisphere?

Day 4

- What imaginary parallel lines run east and west around the globe? What is the name given to the 0° line of latitude that divides the earth into northern and southern hemispheres?
- What is the center point from which all meridians begin and which is the north extremity of the earth's axis? What point on earth is the only place from which all directions are south? (Think about this.)
- What is the parallel of latitude that circumscribes the polar tundra zone: Tropic of Cancer, Arctic Circle, or Antarctic Circle? Approximately what degrees north of the equator is this parallel?

World

Mapping

World (OMB 49)

- Outline the borders of each of the seven continents. Label with BLACK and print with all CAPS.
- Mark the equator with a dotted line. Label with 0° in the left and right margins (borders) of the map. Write "Equator" on the dotted line.
- Locate the Arctic Circle at 66.33°N and the Antarctic Circle at 66.33°S; mark with a dot-dash line and label.

Trail Blazing

Learn how to use an atlas by reading through the informational section in your atlas. Become familiar with your atlas: its glossary, index, and thematic maps.

The globe is the most accurate representation of the earth. We use maps because they are more practical and can be reproduced in a wide range of scales to be effective. Imagine carrying around a globe large enough to read the cities of Europe!

It's important to understand that continent shapes are always distorted on flat maps (TUG 31). To demonstrate the distortion you can use a balloon.

1. On stiff paper draw and cut out a triangle, square, and circle (about two inches wide).
2. Blow up the balloon but do not tie it. Let the balloon represent the shape of the earth.
3. Trace the circle, triangle, and square on the balloon with a felt-tip marker. Let the drawings represent the continents.
4. Let the air out of the balloon.
5. Snip the tip off, and cut the balloon to open out flat, being careful not to cut across your drawings.
6. Pin the corners and sides to a bulletin board or cardboard to form a rectangle.

Compare the shapes on the balloon to the templates you used to trace them. Just like the shape of thefigures changed, drawings of the earth on a map have some level of distortion. A variety of map projections attempt to overcome distortion in one way or another. Maps are very useful for understanding the earth and learning geography. To learn more about map projections, log on to:
http://www.colorado.edu/geography/gcraft/notes/mapproj/mapproj_f.html

Longitude and latitude lines form a grid on maps and help you find places. Read about how to use these grids and play one of the grid games (TUG 124-126).

Learn about planet Earth and the solar system. Study what factors create the seasons.

Geography Notebook

Start a geography notebook (TUG 159). Use some reproducible pages to get you started (TUG 154, 155). Divide into sections by continent. Place map work at the beginning of each section. Plan to add to this notebook weekly.

Make a chart of world facts. Include the circumference and diameter of the earth, the distance between the earth and the sun, the distance between the earth and the moon, the highest point on earth, the lowest point, the highest temperature ever recorded and where, an d the lowest temperature ever recorded and where. Find this information in the *Answer Atlas*, Internet search, or an almanac.

Geography Through Art

- Pangea Puzzle
- Make a World Map
- Make a treasure map and age it (GTA 23-24).

GEOGRAPHY TRAILS

Week 2 - World

Day 1

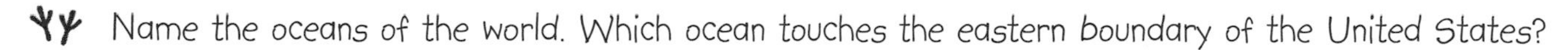

Name the oceans of the world. Which ocean touches the eastern boundary of the United States?

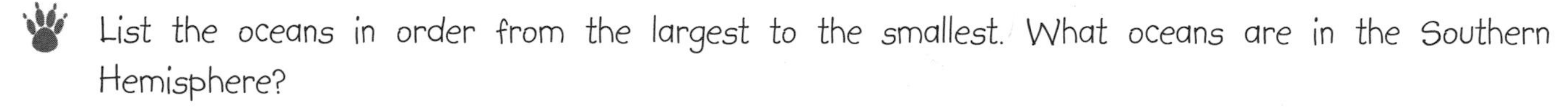

List the oceans in order from the largest to the smallest. What oceans are in the Southern Hemisphere?

List the oceans (with area) in order from largest to smallest. What is the approximate total water volume of all of the oceans?

Day 2

What imaginary lines run north and south on maps and globes? What is the name of the 0° line of longitude that divides the earth into eastern and western hemispheres?

What continents are located in the Eastern Hemisphere? What continents are in the Western Hemisphere?

Through what city does the starting point of lines of longitude pass? What percentage of the earth's surface is covered by water, and what percentage of the earth's water is salt water?

Day 3

What symbol on a map represents a river? What color is often used on a map to represent water?

How much of the water on Earth is fresh water? What kind of map depicts countries, states, provinces, territories, and cities?

What theory states that the earth is divided into large crustal slabs (or plates), constantly in motion? What often occurs when these plates collide or grind past each other?

Day 4

What oceans touch the border of the continent of Australia? What continent is closest to Antarctica?

What oceans touch the border of the continent of North America? What kind of map depicts different land elevations and often represents ocean depths by different colors?

What layers make up the earth, and approximately how thick is each layer? What is the equatorial circumference of the Earth, rounded to the nearest hundred?

Note on oceans:
Many still consider there to be four oceans of the world, however in 2000 the International Hydrographic Organizations (IHO) decided to delimit a fifth ocean. The Southern Ocean is located south of 60°S latitude and surrounds the continent of Antarctica. This body of water was previously the southern part of the Indian Ocean, Atlantic Ocean, and Pacific Ocean.

POINTS OF INTEREST

World

Mapping

World (OMB 49)
- Label the oceans. Use BLUE and print with all CAPS.
- Label the prime meridian at 0° and the international date line at 180°.

Pacific Rim (OMB 40)
🐾 Label the Pacific Ring of Fire on a map of the Pacific rim.

Trail Blazing

Study the composition of the earth. There is a great series of charts depicting the composition of each layer of the earth in the Answer Atlas and in the Usborne Encyclopedia of World Geography. Make your own chart on poster board.

Learn about Pangea and the theory of plate tectonics. Define continental drift.

Volcanoes formed many of the earth's mountains and islands. The earth has over 500 active volcanoes including many on the ocean floor. Study volcanoes. Make your own exploding gas volcano or a volcano notebook (TUG 114-120). Log on to http://volcano.und.nodak.edu, to learn more about volcanoes and the Pacific Ring of Fire.

Study oceans. Look for a world terrain or ocean floor profile diagram in an atlas. If your atlas includes these diagrams, take some time to "see" the oceans and land from a whole different perspective than standard maps provide. Very interesting!

Copy "Good Stuff to Know by Heart" cards and start memorizing these world facts. Practice with friends and family (TUG 281-283).

The International Hydrographic Organization (IHO) has declared and demarcated a fifth ocean and named it the Southern Ocean. It includes all the water below 60°S to the boundary of Antarctica. Learn more about the fifth ocean and how it came to be established. Try using about.com or the CIA Factbook online at: http://www.cia.gov/cia/publications/factbook/geos/oo.html. (Do this week or in Week 26.)

Geography Notebook
Begin to draw the flags of the world. Add new flags each week from countries on the continent that you are studying. Learn what the symbols and colors mean.

Learn the about the layers of the earth, draw a diagram, and label (TUG 116, atlas).

Illustrated Geography Dictionary
Start your own Illustrated Geography Dictionary. Add new words and terms throughout the year. Draw or attach pictures of geographic features and write a simple definition (TUG 339-342). Use the template provided in the appendix. Need help to draw the features? Try using an illustrated geographical terms chart. Start with these terms:

latitude (parallel) hemisphere

Geography Through Art
- Earth from Space
- Build a papier-mâché globe.

GEOGRAPHY TRAILS

Week 3 - North America

Day 1

In what hemisphere is North America located? What is the largest country in North America?

What are the three largest countries in North America? Which has the largest population?

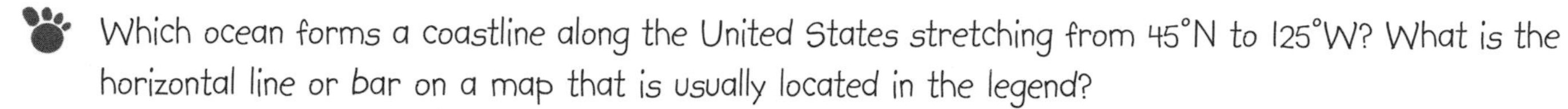
Which ocean forms a coastline along the United States stretching from 45°N to 125°W? What is the horizontal line or bar on a map that is usually located in the legend?

Day 2

What is a small area of land surrounded by water? What three oceans border Canada?

The world's largest island is located in North America and claimed by Denmark. What is its name? What is its area in square miles?

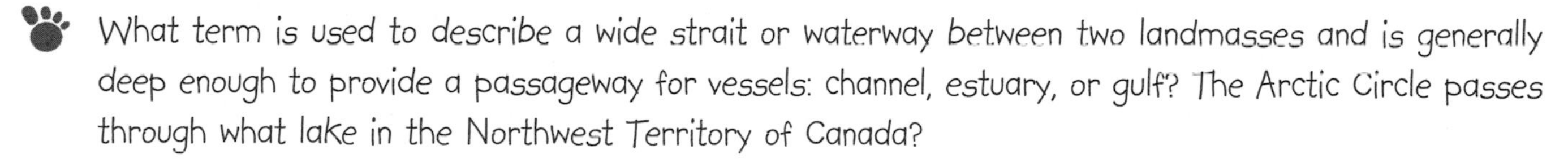
What term is used to describe a wide strait or waterway between two landmasses and is generally deep enough to provide a passageway for vessels: channel, estuary, or gulf? The Arctic Circle passes through what lake in the Northwest Territory of Canada?

Day 3

What countries in North America lie in the Arctic Circle? Is Greenland closer to Mexico or Canada?

What is the name of the newest territory in Canada, which includes the Baffin Islands? What is a peninsula?

Who first used the word "atlas" to describe a collection of maps? What is the capital of the nation whose northern border is the United States?

Day 4

What sea is off the coast of Alaska? What other continent borders this sea?

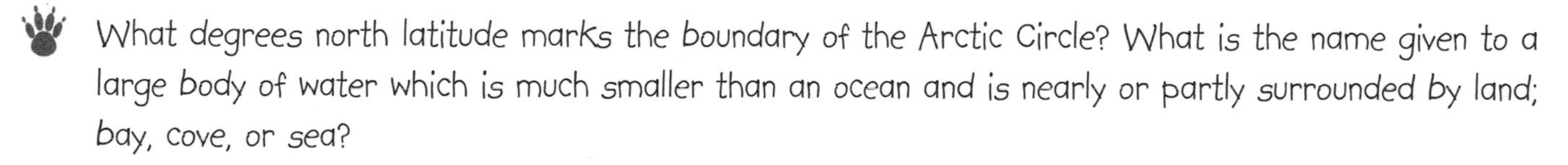
What degrees north latitude marks the boundary of the Arctic Circle? What is the name given to a large body of water which is much smaller than an ocean and is nearly or partly surrounded by land; bay, cove, or sea?

What city, located at 74°W longitude, has the highest population of any in the United States? What percentage of the U.S. population live in urban areas?

POINTS OF INTEREST

North America

Mapping

North America (OMB 39)
- Label Greenland, Mexico, Canada, the United States, and Cuba with black. Use all CAPS.
- Put a brown triangle on Mt. McKinley and label it. Write the elevation under its name.
- Label Arctic Ocean, Pacific Ocean, Atlantic Ocean, and Great Bear Lake in Canada.

OR do North America "Map-It" (TUG 160).

OR do North America Outline Activity (TUG 201-202).

Note: For consistency in map labeling, use uppercase and lowercase letters for city names and other places; use all CAPS when printing continent or country names. Use dot within a circle for capital cities.

Trail Blazing

To enhance your study of physical geography, make a salt dough or clay physical relief map of North America. For recipes see Types of Clay (GTA) and 3-D map instructions (TUG 62). Make a 3-D map of North America. Paint and label.

Read about North America (TUG 160-168, GTA, student atlas) and watch travel videos (from the library) about places in North America.

Learn about the people of North America which includes the U.S., the far north, Mexico, Central America, and the Carribbean.

Learn about Canada's newest territory, Nunavut. Who lives there? When was it established and why? Write a short report or give an oral report of your findings.

Geography Notebook
Start clipping newspaper articles about events occurring in North America. Place in the notebook.

Begin to make your own set of thematic maps of North America. Choose from the following map themes: physical elevations, land usage (environments), population, climate, and natural hazards. Add new themes weekly or when reminded. You should find the information for this project in your student atlas.

1. Make several copies of the North America outline map.
2. Draw and shade the different regions with appropriate colors.
3. Place a legend box in the lower left area of each map.

Illustrated Geography Dictionary
Continue to add new words and terms throughout the year. Use your own drawings, take photographs with a camera, or cut pictures out of magazines. Here are a few more terms to add:

coast (coastline) | channel | sea | harbor

Geography Through Art
- The Eskimos do art rubbings on walrus and whalebone. Try your hand at Scrimshaw.
- Make your own Eskimo Plate.

GEOGRAPHY TRAILS

Week 4 - North America

Day 1

What country is located north of the United States of America? What plain spreads across both countries?

Name four Canadian islands located within the Arctic Circle. What sea is west of Banks Island and north of Alaska?

Descendants of what nationalities make up the majority of Canada's population? What Canadian territories or provinces border the Hudson Bay?

Day 2

What river forms the boundary between the United States and Mexico? What is the main crop of the Great Plains?

What are the two largest countries in North America? Between what two islands is the Davis Strait?

Is the Gulf of St. Lawrence east or west of the St. Lawrence River? What are the names of the three Canadian territories?

Day 3

What ocean forms the western coast of Alaska, Canada, U.S., and Mexico? What countries have land at 135°W?

Into what river does the Ohio River empty? What is the main economic land usage in northern Canada and most of Alaska?

What sound is north of Victoria Island? Name the ten Canadian provinces.

Day 4

Name the large bay in the midst of Canada. Alaska does not touch the boundary of any U.S. state; what country does it touch?

What chain of volcanic islands separates the Bering Sea and the Pacific Ocean? What mountain range forms the terrain of the eastern U.S.?

Name the four U.S. states with volcanic activity in the 1900s. Where are most of Alaska's 34 active (since 1900) volcanoes located?

POINTS OF INTEREST

North America

Mapping

North America (OMB 39)
- Label the Great Plains; shade lightly in brown. They span the central U.S. and Canada.
- Label the Bering Sea.
- Label the Rio Grande River and Mount St. Helens (in Washington state).

Canada (OMB 20)
- Label these places. Use blue for names representing water.

Hudson Bay	Newfoundland Island	Davis Strait
Magnetic North Pole	Lawrence River	Bering Sea
Beaufort Sea	Baffin Island	Niagara Falls

- Label the five Great Lakes; shade lightly in blue.

Trail Blazing

Learn to identify the ten Canadian provinces and three Territories. Make flash cards by drawing the shape of each on one side and writing its name and capital on the reverse. Put a star or circle with a dot on the location of the capital.

Each minute over 15 million cubic feet of water (weighing over 417,00 tons) plunges over Niagara Falls. Learn more about Niagara Falls.

Study the phenomenon of Aurora Borealis. What causes this splendid light show?

Cook a meal typical to the people of Canada from *Eat Your Way Around the World* or from another source.

Learn about animals of the world and start an animal notebook, or put an animal section in the geography notebook. Organize by continent or biomes, beginning with North American wildlife. Add new animals as you learn about each continent (TUG 47).

Geography Notebook

Read about Canada and do "Map-It" (TUG 162). Find travel videos of places located in Canada from the library.

Select 1-3 countries from this continent. Complete the Countries of the World: A Fact Sheet (GTA 171) for each country.

OR Select a country from this continent. Write a short essay from the choices in TUG 218. Include a sketch of the country's flag.

Create a new thematic map for your notebook. See instructions from last week.

Illustrated Geography Dictionary

Continue to add new words and terms throughout the year. Use your own drawings, take photographs with a camera, or cut pictures out of magazines. Here are a few more terms to add:

bay   sound   island   plain

Geography Through Art
- Make your own maple leaf rubbings
- QuickSketch: Moose
- Design your own sand paintings using colored craft sand.

GEOGRAPHY TRAILS

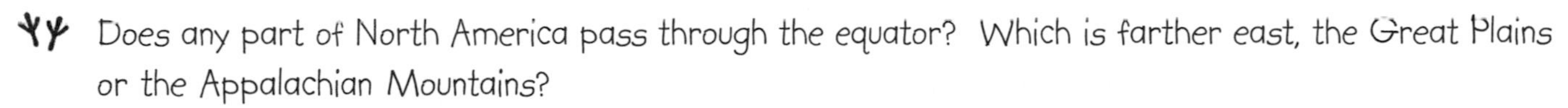

Week 5 - North America

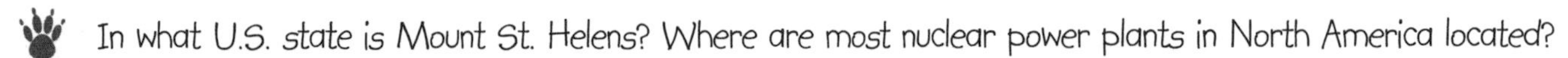

Day 1

- Does any part of North America pass through the equator? Which is farther east, the Great Plains or the Appalachian Mountains?
- In what U.S. state is Mount St. Helens? Where are most nuclear power plants in North America located?
- About 60% of North Americans live in what country? What city is the world's largest and fastest-growing metropolitan area, with a population of over 14 million people?

Day 2

- What is the highest mountain peak (or highest point) in North America? Where is it located?
- Which of the Great Lakes is the deepest? Which one of the Great Lakes is located entirely in the United States?
- What is North America's longest river system? What is the lowest point in North America, where is it, and what is the elevation?

Day 3

- How many lakes make up the Great Lakes? On a map, what symbol is used to depict a mountain peak (or the highest point)?
- What is North America's longest river system? What box with symbols explains the information shown on a map?
- Rivers to the west of this line flow into the Pacific Ocean; rivers to the east flow into the Atlantic; what is its name? What is NAFTA? (hint: use almanac)

Day 4

- Mount Rushmore is in South Dakota; what is the capital? Are the Rocky Mountains east or west of the Mississippi River?
- In what state is the Mississippi River Delta? Name the highest mountain peak in North America and its elevation.
- What is a deep furrow in the ground or the ocean floor: canal, valley, or trench? Using a thematic map, determine what types of landforms make up the majority of the western region of North America.

POINTS OF INTEREST

North America

Mapping

North America (OMB 39)
- Label the Rocky Mountains and Appalachian Mountains.
- Mark the Tropic of Cancer (at 23.5°N) with a dot-dash line and label.
- Draw and label the Mississippi-Missouri Rivers.
- Label Death Valley, North Dakota, Aleutian Islands.
- Label New York City.

Trail Blazing

Learn the countries and capitals of North America. Make flash cards by drawing the shape of each country on one side and writing its name and capital on the reverse. Put a star or circle with a dot on the location of the capital. Or put the name of the state on one side and its capital on the reverse.

Learn about Gutzon Borglum's truly awesome undertaking of carving the solid granite cliff of Mount Rushmore (GTA).

The world's longest bridge is the 24-mile Lake Ponchartrain Causeway in Louisiana. Why was it built? How did it affect the environment, transportation, and the economy?

The world's longest tunnel is the Delaware Tunnel, a 105-mile water tunnel in New York. Learn more about the construction and its effect on the environment, transportation, and the economy.

The Statue of Liberty is over 150 feet tall and was a gift from the French. It stands in New York Harbor welcoming immigrants to settle in the United States. A trip to New York would not be complete without taking the ferry for a close-up view. Learn more about this beautiful American icon (GTA). What is the inscription found at the base of "Lady Liberty"? To take a web tour log on to: http://www.nyctourist.com/liberty 1.htm.

Alaska and Hawaii were the last two territories to officially become states of the United States of America, both in 1959. The U.S. government purchased Alaska from Russia in 1867 and annexed Hawaii in 1898. Learn about the history of these two acquisitions and what made them important to the U.S.

Geography Notebook

Study Native American nations. Map the location of the different nations on a U.S. outline map.

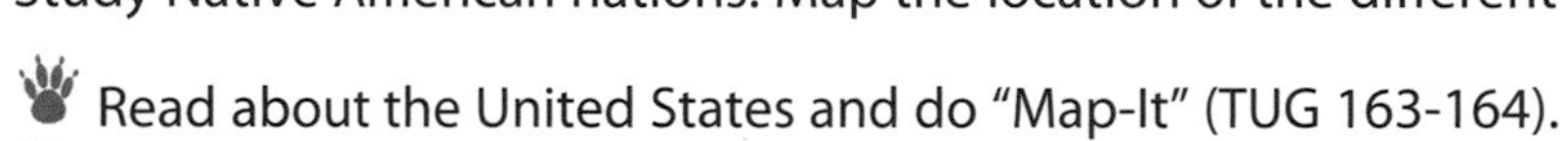

Read about the United States and do "Map-It" (TUG 163-164).

Answer challenge questions for North America (TUG 202).

Learn about any of these fascinating U.S. landmarks: the Arch in St. Louis, Empire State Building in New York, Golden Gate Bridge in San Francisco, and the Seattle Space Needle. Find pictures of these or other landmarks to put in your notebook. Include a map identifying where they are located.

Geography Through Art
- QuickSketch: Mount Rushmore
- Central Park
- Cityscape
- Learn some Indian signs and symbols. *Geography Through Art* has over twenty symbols that you can use. Adapt them to make new symbols. Write a short story using the symbols and see if a friend can "read" your story.

GEOGRAPHY TRAILS

Week 6 - North America

Day 1

- What is the name for a large area of water within a curved coastline or partially surrounded by land? (hint: see geographical terms) Name the large body of water that stretches out along the borders of eastern Mexico and southern U.S.
- Name the four largest Caribbean countries. What Caribbean island is located nearest the Tropic of Cancer?
- What channel runs between Cuba and Mexico? What are the names of the two groups of islands in the West Indies that separate the Caribbean Sea from the Atlantic Ocean?

Day 2

- What river carries water from the Missouri River and the Ohio River and empties into the Gulf of Mexico? Name the capital of Mexico.
- Name three peninsulas in North America between 20° and 30° north latitude. What country shares a boundary with both Honduras and Costa Rica?
- Does more land in Mexico rise below 1000 feet or above 1000 feet? In what part of Mexico does the Sierra Madre Occidental mountain range lie?

Day 3

- What river forms the southern boundary of the United States? What term is used for a drawing of the earth's surface?
- What strip of water provides a passageway between Mexico and Cuba? What five states in the U.S. border the Gulf of Mexico?
- What trench is located near 20°N, 65°W? Which has a higher elevation: Guadalajara, Mexico or Tegucigalpa, Honduras?

Day 4

- Through what country in North America does the Tropic of Cancer pass? Is this country north or south of Guatemala?
- Which city does not lie near the Mexican border: San Diego, CA; El Paso, TX; Las Cruces, NM; San Antonio, TX? Which city is nearest Tijuana, Mexico?ç
- What is a large high land area that is generally flat: precipice, plateau, or piedmont? What two imaginary lines mark the tropical climate regions?

POINTS OF INTEREST

Mexico

Mapping

North America (OMB 39)

- Shade and label these peninsulas: Baja California, Florida, and Yucatan Peninsula.
- Put a circle with a dot on Mexico City and Washington, D.C., and label. Choose your own color, but remember to use this same color any time you label a capital city.
- Label these principal cities of North America: Los Angeles, CA; Chicago, IL; Philadelphia, PA; and Toronto, Canada.
- Label your home state, state capital, and your home town.

Trail Blazing

Study the ancient city of Chichen Itza, located on the Yucatan peninsula. You can still see the remains of Mayan stone pyramids and temples there, built nearly 1500 years ago. Write an essay or summary of your study. Add pictures or drawings of Mayan art style.

Mexico has several active volcanoes such as Colima and Popocateptl. Find out what happens when they erupt, how the population is affected, and how the country has recovered since the most recent serious eruptions. To help you get started, log on to http://www.latinamericanstudies.org/mexico-volcanoes.htm.

Popular Mexican foods include tortillas (corn pancakes) and tamales. Have a Mexican lunch or dinner this week and learn something new about the people of Mexico.

Learn about the plants that thrive in the climate and soil of Mexico.

Geography Notebook

Read about Mexico and do the "Map-It" (TUG 165-166) on a separate map.

A couple of interesting landmarks include the Pyramid of the Moon in Teotihuacan, Mexico (built in 900-1100) and Anasazi Pueblo in Bonito, New Mexico (built around 1050). Find pictures of these or other landmarks to put in your notebook. Include a map identifying where they are located.

Make a chart of North America facts. Choose ten countries in North America. Place country names down the left side of the chart. Add columns across the top and choose labels from these topics: capital, area, currency, language, principal religion, and natural resources. See how many cells in the chart you can fill in from your resources, the Internet, or an almanac.

Illustrated Geography Dictionary

gulf　　trench　　river

Geography Through Art

Mexicans make colorful layered tissue-paper designs, called Papel Picado, to hang in the window for festive occasions . Make a Papel Picado to hang in your window. Use the colors of Mexico's flag or choose your own colors. You may want to try your hand at any of these additional projects:

- Design and make your own Folk Art Wall Hanging with burlap.
- Oaxacan Woodcarving
- God's Eye

GEOGRAPHY TRAILS

Week 7 - North America

Day 1

- What country in Central America (see note below) borders both Mexico and the Pacific Ocean? What countries south of Mexico have coasts on both the Caribbean Sea and the Pacific Ocean?
- What countries make up the North American region of Central America? What is the name given to the group of islands that enclose the Caribbean Sea and includes Greater Antilles, Lesser Antilles, and the Bahamas?
- What are the boundaries of Central America? Which Central America countries have tropical and subtropical forests?

Day 2

- What countries border both Guatemala and the Pacific Ocean? What canal in Central America links the Atlantic and Pacific Oceans?
- What is the passageway of water that connects the Atlantic Ocean and the Gulf of Mexico? What two countries are located on the southern border of Honduras?
- What isthmus connects North and South America? What long, wide body of water connects two large bodies of water or separates an island from a larger body of land: sound, spit, or isthmus?

Day 3

- What sea is east of Nicaragua? Name the country on the southern border of Nicaragua.
- What capital city is located near 20°S, 100°W? The Panama Canal connects which two oceans?
- What Caribbean island has a higher standard of living because of its large oil reserves, and about how much oil does it export? What term defines a small strip of water that provides an opening between a smaller and larger body of water and may reach from a sea or lake into the shore land: isthmus, inlet, or delta?

Day 4

- In which direction would you travel if you flew from Panama to Cuba? What is the capital of Cuba?
- What island nation is south of Cuba? What is the climate of Belize?
- What is the largest inland body of water in Central America? What is the name for the region that includes Central America, Mexico, and the Caribbean islands?

POINTS OF INTEREST

Central America

Mapping

Central America (OMB 22)

- Label each of the five U.S. states that are on the Gulf of Mexico.
- Label each Central American country and capital.
- Label the Caribbean island countries of Cuba, Haiti, Dominican Republic, and Jamaica.
- Label Puerto Rico.
- Mark the Tropic of Cancer with a dot-dash line and label.
- Write the words "Lesser Antilles" and "Greater Antilles" along the appropriate chains of islands.
- Shade the country of Panama with purple, and draw the Panama Canal in orange.
- Label the Caribbean Sea, Atlantic Ocean, and Gulf of Mexico in blue.

Trail Blazing

Central America is in the continent of North America. Learn the countries and capitals of Central America. (To help you remember the countries in order use this silly sentence: Great big elephants hide near city parks.) Make flash cards by drawing the shape of each country on one side and writing its name and capital on the reverse. Put a star or circle with a dot on the location of the capital.

Find travel videos of places located in Central America at the library. Learn about the land and people of Central America.

Make a jigsaw map. Color an outline map of North America with different colors for each country. Cover with contact paper. Cut into different shapes. See how fast you can put the puzzle together. Or make a large puzzle with card stock (TUG 62).

The Panama Canal opened in 1914 and shortened the journey from the Atlantic to the Pacific by over 5000 miles. Before its opening, ships had to travel around South America. The canal is 50 miles long. Today, many large ocean-going vessels take the trip around Cape Horn because the canal is too narrow. The history of the construction of the canal is fascinating. Learn more about the Panama Canal and write a report. There are a number of good websites with loads of information and great pictures.

Geography Notebook

Read about Central America and do "Map-It" on a separate map (TUG 166-168).

Some gems found in North America include turquoise, aquamarine, jade, agate, and peridot. Find where these are mined and shade the regions on a separate North America map. Research one or two gems to learn more about how they are mined, how they are used, and the effect on the economy.

Illustrated Geography Dictionary

canal   cape

Geography Through Art

- Women in Panama design beautiful blouses with an applique and fabric technique. These special Mola blouses are part of their traditional Kuna dress. Try your hand at copying a Mola design using poster board and brightly colored yarn and cloth.
- Learn about Hupile weavings of Guatemala.
- QuickSketch: Sloth

GEOGRAPHY TRAILS

Week 8 - South America

Day 1

- What is the largest country in South America? What does a star or dot inside a circle usually represent on a map?
- Name the South American countries counterclockwise in order, starting with French Guiana and ending with Brazil. Which is largest?
- The world's highest waterfall is located in a remote Venezuelan forest; what is its name and how far does it drop? What type of climate covers most of the northern half of the continent?

Day 2

- Two countries in South America produce more coffee than any other countries in the world; name them. Which of these coffee producing countries border Central America?
- Do the Andes Mountains stretch north and south across South America or east and west? What term refers to a group of islands: archipelago, isthmus, or shoal?
- What river delta is near the islands of Trinidad and Tobago? What mountain chain is the longest in the world?

Day 3

- Are the Andes Mountains on the eastern or western part of South America? What two countries have borders along the Andes Mountain range?
- How much rain must a forest receive to be called a rain forest? What islands are located east of the tip of South America, and who governs them?
- Where is the Patagonia region? The highest point, lowest point, hottest place, and coldest place in South America are all in what country? Name these places.

Day 4

- What two South American countries have no coastline? The Incas left stone ruins called Machu Picchu in the mountains of what country?
- What empire controlled the Andes Mountains and the Pacific Coast at the time of Columbus's discovery? What is the name of the archipelago located at the equator, west of Ecuador?
- What southeast city of Brazil is located near 23°N, 43°W? What desert in northern Chile includes areas where no rainfall has ever been recorded and is one of the driest places on earth?

POINTS OF INTEREST

South America, Andes

Mapping

South America (OMB 45)

- Label the Andes Mountain range (shade with light brown).
- Label each country. Use all CAPS.
- Put a circle with a dot on each capital city and label with the city name. Remember to use the same color you have already used to label capital cities.
- Label the Galapagos Islands, Falkland Islands.
- Shade French Guiana. It is not officially a country but a territory of France.

OR do South America "Map-It" (TUG 170).

OR do South America Outline Activity (TUG 203).

Trail Blazing

Make a salt dough or clay physical relief map of South America.

Read about South America; watch travel videos (from the library) of places located in South America (TUG 169-170, GTA, student atlas), and learn about the people of South America.

Cook a meal typical to the people of Peru or Brazil from *Eat Your Way Around the World* or from another source.

Make a jigsaw map. Color an outline map of South America using instructions from last week. See how fast you can put the puzzle together.

Learn about the culture of the people of the Andes. For pictures, songs, stories, and more log on to: http://www.andes.org.

On February 21, 1996, a large earthquake struck about 75 miles off the northern coast of Peru. The earthquake created a tsunami that reached the city of Chimbote, Peru. Learn more about earthquakes and their contribution to formation of tsunamis, or find out about other natural hazards that people of South America have faced.

Geography Notebook

Answer challenge questions for South America (TUG 204).

Learn about animals of South America and add to the animal notebook or the animal section in the geography notebook. Remember to add new animals as you learn about each continent. Here are a few that are fascinating to study: anaconda, Galapagos Tortoise, iguana, macaw, toucan, flamingo, giant armadillo, giant anteater, two-toed sloth, and llama. How do the climate and terrain contribute to the animals' survival?

Make a chart of South America facts. Select six countries in South America for this week and six countries next week. Follow instructions from Week 6.

Make a crossword puzzle with the South American countries as clues and their capitals as the answer. Make copies to share with someone else.

Watch for and clip newspaper articles about events occurring in South America.

Illustrated Geography Dictionary

archipelago　　delta　　savanna(h)

Geography Through Art

- Retablo Art
- QuickSketch: Machu Picchu
- Macaw
- Bulls of Pucura
- Llama

GEOGRAPHY TRAILS

Week 9 - South America

Day 1

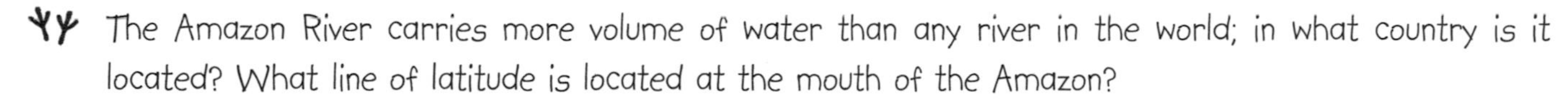
The Amazon River carries more volume of water than any river in the world; in what country is it located? What line of latitude is located at the mouth of the Amazon?

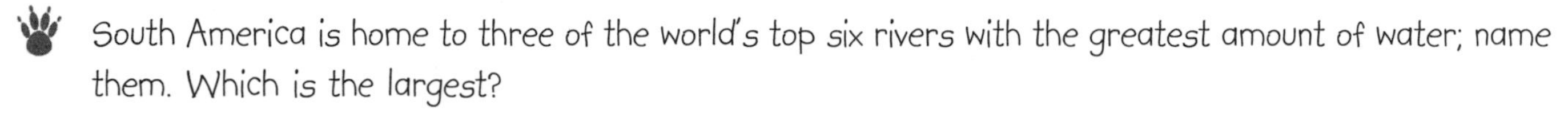
South America is home to three of the world's top six rivers with the greatest amount of water; name them. Which is the largest?

What city, in Argentina's Tierra del Fuego province, is the southernmost city in the world? What is its latitude?

Day 2

What is the capital of Chile? Is the Caribbean Sea located north or south of South America?

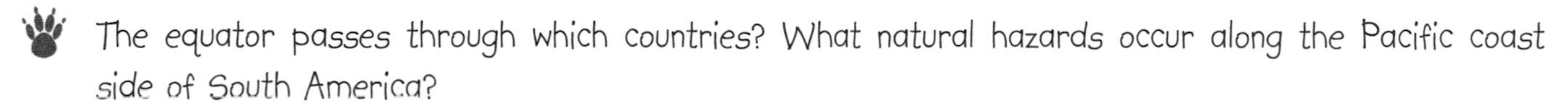
The equator passes through which countries? What natural hazards occur along the Pacific coast side of South America?

What is the name of the cape at the southern tip of South America? How many cubic feet per second flow from the mouth of the Amazon River?

Day 3

What kind of climate describes the northern half of Brazil? Is the majority of South America above or below the equator?

Define tributary. Which country has the most tributaries?

The Andes Mountains pass through what seven countries? Briefly describe the climate in Bogota, Colombia.

Day 4

Which country shares its border with the most South American countries? What is the capital of Venezuela?

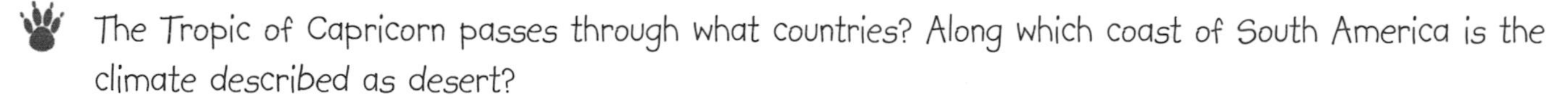
The Tropic of Capricorn passes through what countries? Along which coast of South America is the climate described as desert?

What country is the world's leading exporter of bananas? What two South American countries are landlocked?

POINTS OF INTEREST

South America, Amazon

Mapping

South America (OMB 45)

- Draw and label the Amazon, Orinoco, Parana, 🐾🐾 Negro, and Uruguay rivers.
- Mark the Tropic of Capricorn at 23.5° S with a dot-dash line and label it with blue.
- Label Cape Horn; Caracas, Venezuela; and 🐾🐾 Lake Titicaca.

Trail Blazing

Learn the countries and capitals of South America. Make flash cards by drawing the shape of each country on one side and writing its name and capital on the reverse. Put a star or circle with a dot on the location of the capital.

The world's largest rain forest, the Amazon Rain Forest, covers 2.5 million square miles and takes up two-fifths of South America. Its average temperature is over 80° and average humidity is over 80%. The Amazon River stretches over 4000 miles long and carries 20% of the world's fresh water. It is so big that its mouth is about 200 miles wide! Learn about the plant and animal life of the Amazon. Check out travel videos or National Geographic videos from the library to see this beautiful natural wonder.

Cook a meal typical to the people of Argentina, or Venezuela from *Eat Your Way Around the World* or from another source.

The world's largest waterpower project is the Itaipu Hydroelectric Dam on the Parana River in Brazil and Paraguay. Learn about dams. Take note of construction techniques, choosing a location for a dam, and the impact a dam has upon the ecosystem.

The tallest building in South America is only 656 ft high, less than half the height of the Sears Tower in Chicago. It is the Parque Central Torre Officinas in Caracas, Venezuela. Research other interesting places in and around Caracas and make a travel brochure.

Geography Notebook

Make your own set of thematic maps of South America. Follow the instructions given in Week 3.

Several interesting landmarks from the past include Royal Road of the Incas in the Andes Mountains and Machu Picchu in Peru, both built 1200-1500, and Chan Chan in Peru, built 1000-1400. Find pictures of these to put in your notebook. Include a map identifying where they are located.

Continue making a chart of South America facts on the remaining six countries in South America.

Learn about the principal crops of South America. How do the climate and soil contribute to agriculture?

Select 1-3 countries from this continent. Complete the Countries Fact Sheet (GTA 171) for each.

🐾 OR Select a country from this continent. Write a short essay from the choices in TUG 218. Include a sketch of the country's flag.

Geography Through Art

- Rain Stick
- QuickSketch: Hercules Beetle
- QuickSketch: Bombardier Beetle
- Butterfly Watercolors
- QuickSketch: Frog
- Piñata
- Tagua Nut Carvings

GEOGRAPHY TRAILS

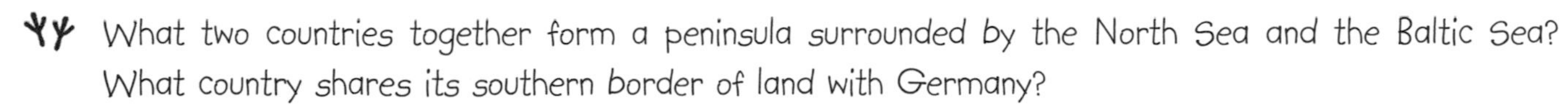

Week 10 - Europe

Day 1

- What two countries together form a peninsula surrounded by the North Sea and the Baltic Sea? What country shares its southern border of land with Germany?
- What countries are on the North Sea? Name the three largest rivers that run through Germany.
- What is the name for the area comprised of Denmark, Norway, Sweden, Finland, and Iceland? What country lies west of the nation whose capital is Berlin and north of the nation whose capital is Brussels?

Day 2

- At what degrees longitude is the prime meridian? Northern Ireland, Scotland, England, and Wales are not really separate countries but together make up what nation?
- What two bodies of water does the Strait of Dover connect? What four countries lie at least partially within the Arctic Circle?
- What is the region above the Arctic Circle that extends over northern parts of Norway, Sweden, Finland, and northwest Russia? What city is the site of international courts of arbitration and justice?

Day 3

- What river flows through southern Germany? What country on the North Sea has Amsterdam for its capital?
- Name the four divisions of the United Kingdom? What channel flows between Wales and Ireland?
- What part of Germany has a higher elevation: the northern or southern? What sea forms a border of Sweden?

Day 4

- What channel passes between France and the United Kingdom? What European island nation is located in the Atlantic Ocean just south of the Arctic Circle?
- What type of climate does western Europe have? What is the main economy of Ireland?
- What countries comprise the British Isles? What principalities form the United Kingdom? (hint: see almanac)

POINTS OF INTEREST

Northern Europe, Germany, Netherlands

Mapping

Scandinavia (OMB 43)
- Label these bodies of water in blue: Barents Sea, Gulf of Finland, Norwegian Sea, Baltic Sea, and Gulf of Bothnia.
- Shade Russia; Draw the Arctic Circle with a dot-dash line and label.
- Label these countries and their capitals: Norway, Sweden, Finland, and Denmark.

Europe (OMB 25)
- Label these countries and their capitals: Iceland, Norway, Sweden, Denmark, Germany, United Kingdom, Ireland, and Russia.
- Color Luxembourg red; shade Netherlands yellow and Belgium orange. Include a legend indicating what colors represent what countries.
- Label these bodies of water: North Sea, Atlantic Ocean, Baltic Sea, English Channel, Danube River, and Strait of Dover.

OR do Europe "Map-It" (TUG 175).

OR do Europe Outline Activity (TUG 205).

Trail Blazing

Make a salt dough or clay physical relief map of Europe. For recipes see Types of Clay (GTA) and 3-D map instructions (TUG 62). Salt dough recipe is given in the instructions on page 12.

Read about Europe (TUG 173-175, student atlas), and find travel videos of places located in Europe.

The Vikings were mostly Icelanders, Norwegians, or Scandinavians. They were great tradesmen and craftsmen. They were great storytellers and passed along their rich history through legends. Study Vikings.

Cook a meal typical to the people of Ireland, Germany, Great Britain, or Netherlands from *Eat Your Way Around the World* (or from another source), and have a taste-testing party.

Active volcanoes have erupted in Iceland in the past 25 years. Find out what happened, how the population was affected, and how the country has recovered (Surtsey, Iceland in 1987; Loki, Iceland, 1996).

Learn about the 4,626-foot suspension bridge, the Humber Bridge, in England; the world's longest undersea tunnel, the Channel Tunnel, connecting England and France; or Big Ben, a famous landmark in London, England. Use an encyclopedia, library books, or the Internet.

Geography Notebook

Start a chart of Europe facts. Follow the instructions from Week 6 choosing ten countries each week for the next three weeks.

Read about the United Kingdom and do the "Map-It" (TUG 176) on a separate map of the British Isles (OMB 18).

Geography Through Art
- Read about the Vikings' rune stones and try your hand at carving them.
- Ivory Chessman - make your own with modeling material.
- Sculpt your own version of Stonehenge, the famous antique structure in Wiltshire, England.
- The story of Hansel and Gretel was set in the Black Forest of Germany. Make your own gingerbread house.
- QuickSketch: Big Ben

GEOGRAPHY TRAILS

Week 11 - Europe

Day 1

- What large mountain range forms the eastern boundary of Europe? What city in Russia is Europe's largest city?
- Through what countries do the Northern European Plains fall? What mountain ranges form the eastern and southern boundary between Europe and Asia?
- What is the capital of the country called the "breadbasket of Europe"? What are two former names for St. Petersburg, Russia?

Day 2

- What three Baltic countries are sandwiched between land in Russia? What country is north of Ukraine and east of Poland?
- In what country are both the Carpathian Mountains and Transylvanian Alps? The Black Sea and the Caspian Sea form part of the border between Europe and Asia; what European countries are on the Black Sea?
- Which country was NOT part of the former Soviet Union: Belarus, Ukraine, or Romania? Which country has mountains: Latvia, Belarus, or Bulgaria?

Day 3

- What body of water forms the border of Ukraine, Romania, and Bulgaria? Is the land in Romania mostly mountainous or mostly plains?
- What country is landlocked by Ukraine and Romania? In what direction would you be going if you traveled from Bulgaria to Estonia?
- What mountain ranges and bodies of water form the boundary between Europe and Asia? Through what three countries lie the Carpathian Mountains?

Day 4

- What river flows along the border of Romania and Bulgaria? What river starts north of Moscow and empties into the Caspian Sea?
- Which has a higher elevation: Helsinki, Finland, or Sofia, Bulgaria? What is the capital of the country that is south of Latvia and has coastline on the Baltic Sea?
- What is the capital of the Scandinavian nation whose eastern border is Russia? Moldova is landlocked between what two countries?

POINTS OF INTEREST

Eastern Europe

Mapping

Eastern Europe (OMB 24)
- Label these countries and their capitals: Estonia, Latvia, Lithuania, Belarus, Ukraine, Romania, Russia Moldova, and Bulgaria.
- Identify the Black Sea and label in blue.

Europe with no boundaries (OMB 26)
- Draw and label these rivers: Danube, Elbe, Rhine (Rhein), Thames, and Volga.
- Draw and label these mountain ranges: Ural, Caucasus, Carpathians Alps, and Apennines.

Trail Blazing

Start learning the countries and capitals of Europe. Make flash cards by drawing the shape of each country on one side and writing its name and capital on the reverse. Follow the instructions from Week 4.

Romanians eat a cornbread mush called mamaliga. Eat something made with cornmeal or try making your own mamaliga. Or, cook a meal typical to the people of Russia from *Eat Your Way Around the World* or from another source.

Learn about one of Europe's many castles. Find out why it was built, if it is still used today, and for what purpose it is used.

Learn about the people of Europe and their diverse cultures.

Geography Notebook

Start your own set of thematic maps of Europe. Follow the instructions given in Week 3. Do two maps this week and two next week. You should find this information in your student atlas.

Select 1-3 countries from Europe. Complete the Countries of the World: A Fact Sheet (GTA 171) for each country. Or design your own sheet that includes area, population, capital, type of government, language, currency, chief cities, natural resources, products, the flag, description of the land, rivers and bodies of water, and a picture or drawing. Add additional countries as you wish during the next two weeks.

Select a country from this continent. Write a short essay from the choices in TUG 218. Feel free to include a sketch of the country's flag.

Add to the chart of Europe facts begun last week. Choose ten new countries this week and continue for the next two weeks.

Answer challenge questions for Europe (TUG 207).

Geography Through Art

Note: Russia is located in both Europe and Asia. It is covered in both Europe and Asia sections in the *Trail Guide to World Geography* but only in Asia in *Geography Through Art*.
- Have you ever seen Matryoshka dolls? They are a Russian folk art of nested dolls of descending size. Learn about them and make your own.
- Russian Pins
- Russian Lubok
- Castles
- Stonehenge
- Russian Mosaics
- Ukrainian Easter Eggs
- Coat of Arms
- Inn Signs

GEOGRAPHY TRAILS

Week 12 - Europe

Day 1

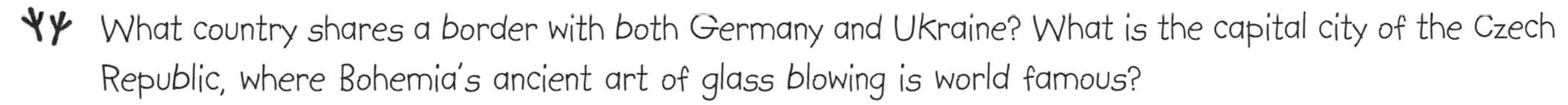

What country shares a border with both Germany and Ukraine? What is the capital city of the Czech Republic, where Bohemia's ancient art of glass blowing is world famous?

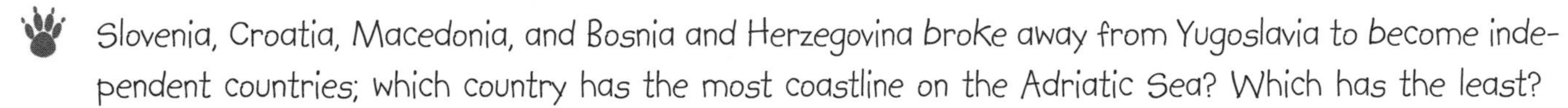

Slovenia, Croatia, Macedonia, and Bosnia and Herzegovina broke away from Yugoslavia to become independent countries; which country has the most coastline on the Adriatic Sea? Which has the least?

What capital is on the Vistula River? In what direction would you be going if you traveled from Bucharest, Romania, to the Carpathian Mountains?

Day 2

What river flows south through Hungary? The Adriatic Sea forms the eastern boundary of Italy; name two countries on the Adriatic Sea across from Italy?

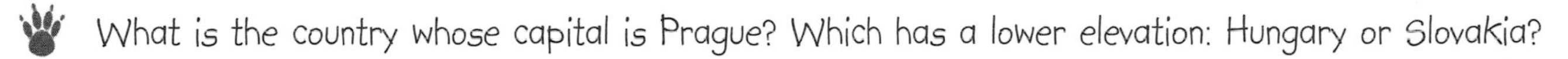

What is the country whose capital is Prague? Which has a lower elevation: Hungary or Slovakia?

Which was NOT once a part of Yugoslavia: Macedonia, Slovenia, Slovakia? What are the capitals of the two countries that were once the nation of Czechoslovakia?

Day 3

If you took a train from Macedonia to Slovakia, in what direction would you be traveling? Name one capital city on the Danube River.

Through what country does the Vistula (Wisla) River flow? Which city has a higher population: Sarajevo or Budapest?

What countries are on the eastern boundary of the Adriatic Sea? Sarajevo is the capital of what country?

Day 4

Is most of Europe in the Eastern or Western Hemisphere? (hint: see handbook section in your atlas) What is name for the part of a map that shows N, E, S, W directions?

What country is Macedonia's western neighbor? In which country is the economy more dependent upon forestry: Hungary, Macedonia or Albania?

Which country has a Mediterranean climate of short rainy winters and long dry summers: Hungary or Albania? Name three capital cities located on tributaries of the Danube River.

POINTS OF INTEREST

Eastern Europe

Mapping

Eastern Europe (OMB 24)
- Add these countries and their capitals to last week's map of Eastern Europe: Poland, Slovakia, Hungary, Serbia, Montenegro (Serbia and Montenegro), Albania, and Macedonia.
- A small part of Turkey is in Europe. Shade it green and label it.

Europe (OMB 25)
- Shade Bosnia and Herzegovina pink. Place a pink shaded box in the legend and write the names in it.
- Shade Croatia purple and identify it in the legend.
- Print "Czech" inside the boundaries of the Czech Republic.

Note: For consistency in map labeling, use uppercase and lowercase letters for city names and other places; use all CAPS when printing continent or country names. Use a dot within a circle for capital cities.

Trail Blazing

In 1991 when the U.S.S.R. (Union of Soviet Socialist Republics) broke up, its 15 republics became independent countries. Study the events leading to the breakup of the Soviet Union. What form of government was the Soviet Union, and what forms of government do each of the independent countries have now? What differences has being independent brought to the people of these nations?

Yugoslavia has experienced major changes in recent history. The land has been divided into several nations including Serbia, Montenegro, and Kosovo (which declared its independence in 2008) and could continue to undergo further changes. Study its transformation from six republics to several independent nations. Share your findings in an oral report or written summary. Include pictures if possible.

Study your flash cards and continue to learn each of Europe's countries and their capitals.

Color an outline map of Europe. Cover with contact paper and cut into fun shapes. See if you can name each country as you put the puzzle pieces together.

Make a crossword puzzle using the countries as the answers and the capitals as clues. Swap crosswords with someone else and time how quickly you can answer.

Get a pen pal from a European country by contacting the National Geographic Society's Pen Pal Network or any other pen pal organization (TUG 74).

Geography Notebook
Add to your set of thematic maps of Europe begun last week. Do two more maps from among these themes: physical elevations, land usage (environments), population, climate, natural hazards.

Add to the chart of Europe facts begun two weeks ago. Choose 10-12 new countries this week.

Learn about animals of Europe and add to the animal notebook or the animal section in the geography notebook. Remember to add new animals as you learn about each continent (TUG 47).

Geography Through Art
- Polish paper cutting

GEOGRAPHY TRAILS

Week 13 - Europe

Day 1

- What two countries south of the Pyrenees mountains form a peninsula? In what city would you find the Eiffel Tower?
- What term is used to identify an inlet of an ocean, partially surrounded by land: gulf, bay, or cove? What two countries have coastline on the Bay of Biscay?
- Where is Europe's Grand Canyon, considered one of the natural wonders of Europe, located? What is the capital of the nation with coastline on the Aegean Sea on the east and the Ionian Sea on the west?

Day 2

- What passageway of water connects the Atlantic Ocean and the Mediterranean Sea south of Spain? In what city would you find the Parthenon?
- What tiny country is entirely surrounded by Italy? What is the natural boundary between France and Spain?
- The smallest country in Europe has an area of only .2 square miles and is surrounded entirely by the city of Rome, Italy: what is this country? If you were at 40°N, 5°E would you be walking or swimming?

Day 3

- What country is a peninsula in the Mediterranean Sea and Adriatic Sea? Name four countries that border Italy.
- What tiny country is located in the Pyrenees Mountains between France and Spain? Sicily is a part of what country?
- The Alps form the border with Italy and what other countries? What country, together with Spain, forms a peninsula with the Atlantic Ocean on the west and the Mediterranean Sea on the east?

Day 4

- What mountain range stretches across the border of Switzerland and Italy? Name a world famous mountain peak in the Alps. (It is over 14,000 feet high and was first scaled in 1865!)
- What mountain range runs along the center of Italy? What mountain erupted in A.D. 79 burying the ancient city of Pompeii, Italy?
- What tiny island is the only European country located entirely in the Mediterranean Sea? What tiny country, located in the Italian peninsula, is the smallest republic in the world and claims to be the oldest state in Europe?

POINTS OF INTEREST

Southern Europe & the Mediterranean

Mapping

Mediterranean Sea (OMB 35)

- Label the following countries and capitals in southern Europe: Portugal, Spain, France, Italy, Switzerland, Austria, and Greece.
- Shade Andorra green and San Marino orange. Indicate the names of these places in a legend.
- Circle the island nation of Malta and label.
- Label the following bodies of water in blue: Mediterranean Sea, Black Sea, Bay of Biscay, Aegean Sea, and Strait of Gibraltar.
- Shade Greece and its islands yellow; Italy and its islands (Sardinia, Sicily) red; France and Corsica purple.

Note: For consistency in map labeling use uppercase and lowercase letters for city names and other places; use all CAPS when printing continent or country names. Use a dot within a circle for capital cities.

Trail Blazing

Study the climate of the Mediterranean region. Places with hot dry summers and moderate rainy winters are commonly referred to as having a Mediterranean climate. Study a climate map of the world. Where else is this climate found? Notice the similarities in latitude north and south of the equator. In what part of the United States would you find a similar climate? Associate the climate with the agriculture.

Practice your memorization of the countries and capitals of Europe with the flash cards. How well do you know them? Find out by labeling them on an outline map from memory.

Cook a meal typical to the people of France, Italy, or Spain from *Eat Your Way Around the World* or from another source.

Learn about the Maginot Line fortress built during WWII to protect France from invading Germany. Did it work? Why or why not? How is it used today? For a good website for this study, log on to: http://www.geocities.com/Athens/Forum/1491/index.html

Geography Notebook

Read about France and do the "Map-It" (TUG 177-178) on a separate map of France (OMB 27).

Several interesting landmarks from the past include the Acropolis in Athens, Greece, begun in the 400s B.C.; Colosseum in Rome, Italy, begun around 70 B.C.; and Stonehenge in England, built 1800-1400 B.C. Find pictures of these or other landmarks to put in your notebook. Include a map identifying where they are located.

Illustrated Geography Dictionary

volcano peninsula strait

Geography Through Art

Europe is rich in the arts and famous for the value the people place on the arts. Art museums abound from the Netherlands to France to Italy. The Renaissance brought art expression to a heightened degree. Select from a variety of art projects and lessons.

- The Parthenon in Athens, Greece (or Nashville, Tennessee!)
- QuickSketch: Leaning Tower of Pisa
- QuickSketch: Arch of Septimius Severus
- Sistine Chapel
- Stained Glass Windows
- Minoan Fresco
- Roman Concrete
- Mona Lisa Portrait
- QuickSketch: Eiffel Tower
- Palace at Versailles

GEOGRAPHY TRAILS

Week 14 - Africa

Day 1

- What body of water forms the northern boundary of the continent of Africa? What mountain range is in Morocco and Algeria?

- What African countries have a coastline on the Mediterranean Sea? What gulf is off the coast of Libya?
- What capital city is its nation's commercial center and transportation hub due to its location at the point where the White Nile, Blue Nile, and the Nile Rivers meet? What is Africa's northernmost point and in what country is it located?

Day 2

- What African country forms a peninsula in the Indian Ocean? What three African countries have a coastline on the Red Sea?

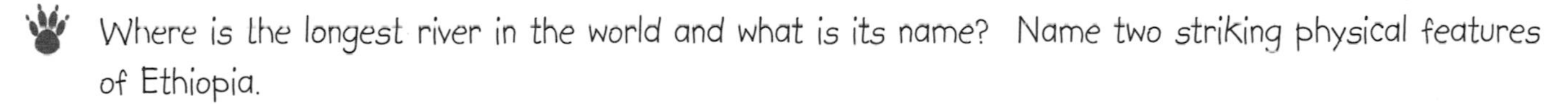

- Where is the longest river in the world and what is its name? Name two striking physical features of Ethiopia.
- What countries make up what is called the Horn of Africa? What natural resource contributes significantly to the economy of Nigeria, Libya, and Algeria?

Day 3

- What is the term for a large land area that receives little rainfall? The Nile is the longest river in the world; through what countries does it flow?
- The continent of Africa has the largest desert in the world, stretching from the Atlantic Ocean to the Red Sea; what is the name of the desert? What countries are located on the Tropic of Cancer?
- In what direction does the Nile River flow? What is the capital of the Mediterranean country west of Egypt and east of Algeria?

Day 4

- What is the world's largest desert? Through which country does the Niger River NOT flow: Chad, Mali, or Niger?
- What is the elevation of Mt. Emi Koussi in Chad? What is the country north of Ethiopia and east of Sudan and what forms its eastern boundary?
- What mountain range stretches across Morocco and northern Algeria? What three landlocked nations have most of their landmass located in the Sahara Desert?

POINTS OF INTEREST

Northern Africa

Mapping

Africa (OMB 7)
Start your own physical map of Africa.
- Label these bodies of water in blue: Mediterranean Sea, Atlantic Ocean, Red Sea, Indian Ocean, and Gulf of Sidra.
- Draw and label the Nile, White Nile, and Blue Nile Rivers.
- Shade the Atlas mountain range with brown and label. Shade the Sahara Desert in yellow and label. Shade the Ethiopian Plateau in orange and label.
- Place a dot-dash line across the Tropic of Cancer and label.

OR do Africa "Map-It" (TUG 191).

OR do Africa Outline Activity (TUG 212-213).

Trail Blazing

Study deserts. Learn what kinds of plants and animals are best suited to desert climate. Using an environment or land-use thematic map of the world, locate all of the world's major deserts. At what degrees latitude above and below the equator do these mostly lie? Now use a world climate thematic map, a world population thematic map, and a world economies thematic map, and look at these same desert areas in each of these maps. What can you conclude from your observation and comparisons of the information on these maps?

Learn about the Great Rift Valley and the role of tectonics in its formation.

Geography Notebook
Obtain two additional outline maps of Africa. You will add to these maps throughout your study of Africa.
1. On the top of one write the title: "Countries of Africa." Label the following countries: Morocco, Algeria, Tunisia, Libya, Egypt, Mauritania, Mali, Niger, Chad, Sudan, Eritrea, Ethiopia, Somalia. Remember to use all CAPS when printing country names.
2. On the top of the second Africa outline map write the title: "Capital Cities." Label it with the capitals of these same countries. Remember to place a dot within a circle at the proper location for each capital city.

Select 1-3 countries from this continent. Complete the Countries of the World: A Fact Sheet (GTA 171) for each country. Follow instructions given in Week 11. Add more countries from Africa during the next three weeks.

Select a country from this continent. Write a short essay from the choices in TUG 218. Include a sketch of the country's flag.

Several interesting landmarks from the past include the Sphinx and Giza Pyramids in Egypt begun around 2500 B.C., and the Temple of Ramses II in Egypt built around 1250 B.C. Find pictures of these or other landmarks to put in your notebook. Include a map identifying where they are located.

Illustrated Geography Dictionary
plateau
desert

Geography Through Art
- Make your own textile design on graph paper from Morroco rug making.
- QuickSketch: Pyramids at Giza
- Egyptian-style drawing
- Hieroglyphics
- QuickSketch: Cats
- Carved Orthostat

GEOGRAPHY TRAILS

Week 15 - Africa

Day 1

- How many African countries have the Atlantic Ocean for their southern boundary? Name three countries with land west of 15°W.
- Name three landlocked countries with land west of 10°E. What country was founded as a refuge for African Americans returning to Africa and is named from the Latin word meaning free?
- What four countries south of 15°N and west of 15°W have coastlines on the Atlantic Ocean? What country has its capital city named for President James Monroe and was founded by freed American slaves?

Day 2

- What landlocked nation forms the northern border of Ghana and Togo? What country has Abuja for its capital?
- What country is bordered by Senegal on three sides and the Atlantic Ocean on the west? What name is given to both an island country in the Atlantic Ocean and a narrow piece of land from Senegal that sticks out into the Atlantic?
- What is the semiarid region that separates the Sahara Desert from the tropical savanna and rain forests of central Africa? What countries have coastline on the Gulf of Guinea?

Day 3

- What country is made up of a group of islands in the Atlantic Ocean west of Mauritania? (hint: see islands near 15°N) The people of Côte d'Ivoire speak French and some African languages. Is Côte d'Ivoire in the eastern or western part of Africa?
- What is the name of the semiarid area just south of the Sahara? Which country does NOT have land with elevations over 1000 ft: Guinea, Côte d'Ivoire, or Guinea-Bissau?
- Where is the Niger Delta? What country has a higher average elevation: Côte d'Ivoire, Sierra Leone, or Guinea?

Day 4

- What are the main usages of most of the land in Africa? Is Liberia east or west of the prime meridian?
- What is the largest river in the western part of Africa? What country is sandwiched between Ghana and Benin?
- What is the capital of the country located north of Ghana and south of Mali? What is the translation of Côte d'Ivoire?

POINTS OF INTEREST

Western Africa

Mapping

Africa Physical Map - add to the map from last week.

- Shade the Sahel region with mustard yellow.
- Shade the higher elevated area in Guinea with light orange.
- Draw and label the Niger River; include the delta.
- Label the Gulf of Guinea.

Trail Blazing

Learn about the Sahel region between the Sahara Desert and the tropical savanna and rain forests.

Start learning the countries and capitals of Africa. Use the Capital Cities map you have begun. See if you can name the country of each capital city you have labeled. Now use the Countries of Africa map and see if you can name the capital for each country labeled.

The Ivory Coast is the English name for the nation of Côte d'Ivoire, officially Republic of Côte d'Ivoire. It was so named because ivory traders frequented the Atlantic coast there in the early days. More recently the people of this nation have adopted its French pronunciation and spelling (try it: coat-dee-'vwar). Learn what life is like there today and make a travel brochure.

Cook a meal typical to the people of Egypt, Ethiopia, or Morocco from *Eat Your Way Around the World* or from another source.

Lake Volta in Ghana is the world's largest artificial lake. Learn about the construction of man-made lakes and the benefit (or damage) to the environment. Or learn more about this modern engineering marvel.

Read about Africa (GTA, TUG 188-190, student atlas), find travel videos from the library of places in Africa and and learn about the people of Africa.

Geography Notebook

Make a chart of Africa facts. Choose ten countries from among those covered the past two days. Follow the instructions given in Week 6. Continue to add to this chart each week for two weeks.

Add more Countries of the World Fact Sheet pages to your notebook with your choice of African nations.

Countries of Africa map: Add the following to the country map you began last week: Senegal, Guinea-Bissau, Guinea, Sierra Leone, Liberia, Côte d'Ivoire, Burkina Faso, Ghana, Togo, Benin, Nigeria.

Capital Cities map: Label the capital of each country listed above.

Start your own set of thematic maps of Africa. Choose from the following map themes: physical elevations, land usage (environments), population, climate, natural hazards. Follow instructions given in Week 3. Do two maps this week and two next week.

Geography Through Art

- African sculpture
- Coffin Art from Ghana
- Golden staff finials
- African masks
- Kente cloth

GEOGRAPHY TRAILS

Week 16 - Africa

Day 1

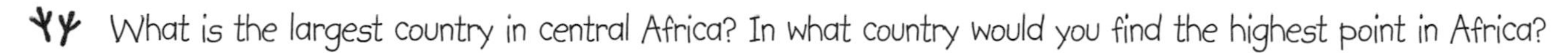

What is the largest country in central Africa? In what country would you find the highest point in Africa?

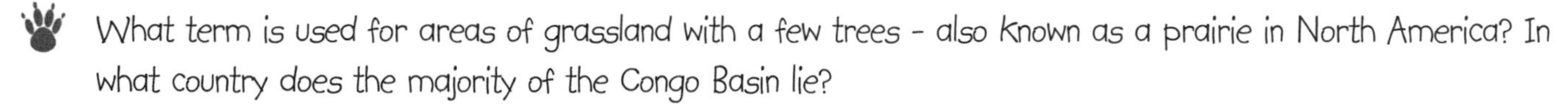

What term is used for areas of grassland with a few trees - also known as a prairie in North America? In what country does the majority of the Congo Basin lie?

In what part of the Democratic Republic of the Congo is the Congo Basin located? Lake Victoria borders what three countries?

Day 2

In what direction would you go to travel from Gabon to Kenya? Traveling at the equator (from Gabon to Kenya), through what countries would you pass?

What is the climate in Central Africa? What large lake in central Africa is the third largest in the world?

What national park contains a great concentration of African wildlife which includes elephants, cheetahs, gazelles, giraffes, hyenas, leopards, zebras, and black rhinoceros? What country is sandwiched between the Congo River Basin and the Sahara Desert?

Day 3

What two countries form the southern border of Chad? What ocean forms the eastern boundary of Africa?

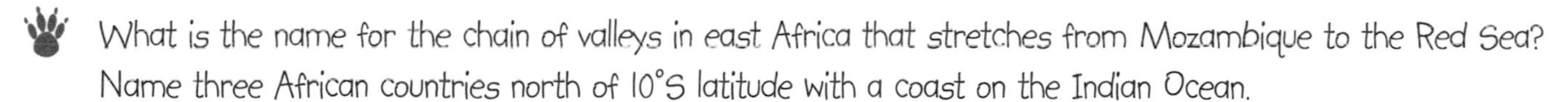

What is the name for the chain of valleys in east Africa that stretches from Mozambique to the Red Sea? Name three African countries north of 10°S latitude with a coast on the Indian Ocean.

What is the vast plain of grassland, acacia bushes, forest, and rocky outcrops located in northern Tanzania? What Atlantic coast country is located at the equator and west of the country of Congo?

Day 4

What river crosses the equator twice and flows into the Atlantic Ocean? Name one animal that lives in rain forests near the equator.

Kilimanjaro is the highest mountain peak in Africa; where is it, and what is its elevation? From a climate map of Africa, name all countries with rain all year.

What is the capital of the country that is home to the tallest mountain peak in Africa? What is the elevation of Mount Karisimbi Volcano, a peak on the boundary between Rwanda and Democratic Republic of the Congo?

POINTS OF INTEREST

Central Africa

Mapping

Africa Physical Map - add to the physical map.

- Place a dotted line across the Equator and label.
- Shade the (Great) Rift Valley in purple and label.
- Label in blue: Lake Victoria, Congo River, Lake Rudolf, and Lake Tanganyika.
- Shade the Serengeti Plain in orange and label.
- Place a triangle at Mount Kilimanjaro and at Cameroon Mountain and label them.
- Shade the higher elevation in Cameroon with light orange.

Trail Blazing

Make a salt dough or clay physical relief map of Africa.

Continue to learn the countries and capitals of Africa. Use the Capital Cities map you have begun. See if you can name the country of each capital city you have labeled. Now use the Countries of Africa map and see if you can name the capital for each country labeled.

Make a jigsaw map from instructions given in Week 7. See how fast you can put the puzzle together.

Cook a meal typical to the people of Kenya or Nigeria from *Eat Your Way Around the World* or from another source.

The beautiful Serengeti region in Tanzania has such a unique ecosystem of climate, vegetation, and fauna that many call it paradise. Over one million wildebeest and over 200,000 zebra migrate through this region several times a year in search of water during the rainy seasons. The Serengeti National Park makes up 14% of Tanzania's land area and serves as a protected area for a wide range of animals - many not found in any other place in the world. Study the Serengeti and learn more about this fascinating region in Tanzania and about the unique migration. Log on to http://www.serengeti.org. (Choose "English", "magnificent wildlife", and "great migration." For an online lesson plan on the wildebeest migration go to: http://www.nationalgeographic.com/xpeditions/lessons/09/gk2/migrationwildebeest.html

Geography Notebook

Start learning about animals of Africa and add to the notebook. Try this website for information on animals in the Serengeti: http://www.serengeti.org/animals.html.

Countries of Africa map: Add these countries to your map: Cameroon, Equatorial Guinea, Gabon, Kenya, Uganda, Central African Republic, Congo, Democratic Republic of the Congo, Rwanda, Burundi, Tanzania.

Capital Cities map: Label the capital of each country listed above.

Select 1-3 more countries from this continent. Complete the Countries of the World: A Fact Sheet (GTA 171) for each country or add ten new countries to the chart of Africa facts begun last week.

Continue watching for and clipping newspaper articles about events occurring in Africa.

Add to your set of thematic maps of Africa begun last week. Do two more maps from these map themes: physical elevations, land usage (environments), population, climate, natural hazards.

Geography Through Art

- Make a collage using "Butterfly Art" from the Central Republic of Africa.
- QuickSketch: Zebra
- Mt. Kilimanjaro, Kenya

GEOGRAPHY TRAILS

Week 17 - Africa

Day 1

- Name two of the landlocked countries at 15°S latitude. What countries are located along the Tropic of Capricorn?
- What country is the world's leading producer of gold? What are the names of two deserts in southern Africa and in what countries are they located?
- Where is the world's largest inland river delta? Where is the Kalahari Desert located?

Day 2

- What cape is at Cape Town, South Africa? What desert is located in Namibia?
- What mountain range extends through Lesotho and the southeastern part of South Africa? What island nation lies east of the Mozambique Channel?
- What country south of the equator has the most land with an elevation below 1000 feet? What channel connects the Indian Ocean at both ends?

Day 3

- What is the main land use for the southern tip of Africa? What Indian Ocean island is east of Mozambique?
- Does the water from Victoria Falls flow into Lake Victoria? What small country lies at the southern tip of Mozambique and east of South Africa?
- What is the capital of the country with the world's largest known deposits of gold, manganese, and platinum? What river forms the boundary between Zambia and Zimbabwe and empties into the Mozambique Channel?

Day 4

- In what parts of Africa does the climate change with the seasons? What river flows through the Democratic Republic of the Congo?
- What country is traversed by the Great Rift Valley and has its land along Lake Nyasa? What kind of climate has the region containing Johannesburg, South Africa?
- What is the name of the arid region along the entire length of the coast of Namibia? What is the country whose capital is Antananarivo?

POINTS OF INTEREST

Southern Africa

Mapping

Africa Physical Map - add to the physical map.

- Place a dot-dash line across the Tropic of Capricorn and label.
- Label in blue: Lake Nyasa, Victoria Falls, Zambezi River, Okavango River, and Mozambique Channel.
- Shade the Namib Desert and Kalahari Desert yellow and label.
- Shade the higher elevation in Madagascar, Mozambique, and South Africa with light orange.
- Label the Drakensberg mountain range.
- Shade all low elevation areas of Africa (with elevation below 200 feet) green.

Trail Blazing

Continue to learn the countries and capitals of Africa. Use the Capital Cities Map you have begun. See if you can name the country of each capital city you have labeled. Now use the Countries of Africa Map and see if you can name the capital for each country labeled.

View video documentaries set in Africa produced by National Geographic, Nova, or any other good source.

Cook a meal typical to the people of South Africa from *Eat Your Way Around the World* or from another source.

Africa is the world's greatest producer of diamonds. The world's largest excavation was the Kimberly Mine in South Africa, over 3000 feet deep and 1500 feet wide. Study diamond mining in South Africa.

Study geology. Learn the difference between igneous, sedimentary, and metamorphic rocks. What is the Mohs hardness scale, and how is it used?

Review world facts from "Good Stuff to Know by Heart" cards (TUG 281-283), or create your own cards with world facts. Include data such as largest islands, largest deserts, largest countries, and world extremes (highest mountain, lowest point, longest rivers, etc.), Practice with friends and family.

Geography Notebook

Add ten new countries to the chart of Africa facts begun two weeks ago.

Answer challenge questions for Africa (TUG 214).

Countries of Africa map: Add Angola, Zambia, Malawi, Namibia, Botswana, Zimbabwe, Mozambique, Madagascar, and South Africa.

Capital Cities map: Label the capital of each country listed above.

Make a crossword puzzle with the African countries as clues and their capitals as the answer. Make copies to share or swap with someone else.

Continue learning about animals of Africa and add to the animal notebook or the animal section in the geography notebook.

Some gems found in Africa include the diamond, emerald, garnet, and tourmaline. Find where these are mined and shade the regions on a separate Africa map. Research one or two gems to learn more about how they are mined, how they are used, and the effect on the economy.

Geography Through Art

- Animal art- learn positive and negative space using animals.
- QuickSketch: Giraffe
- QuickSketch: Lion

GEOGRAPHY TRAILS

Week 18 - Review

Day 1

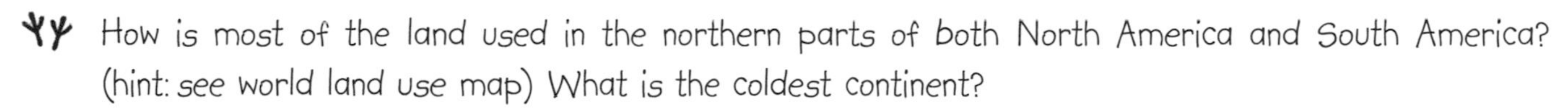
How is most of the land used in the northern parts of both North America and South America? (hint: see world land use map) What is the coldest continent?

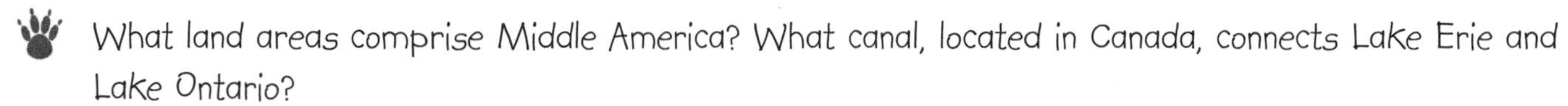
What land areas comprise Middle America? What canal, located in Canada, connects Lake Erie and Lake Ontario?

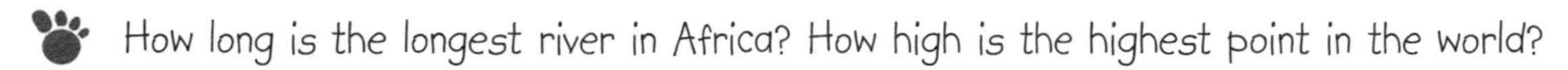
How long is the longest river in Africa? How high is the highest point in the world?

Day 2

What is the largest desert in the world? Is most of Africa east or west of the prime meridian?

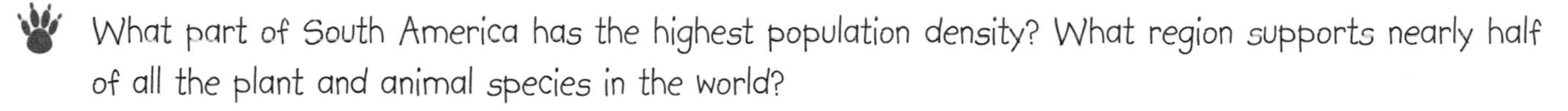
What part of South America has the highest population density? What region supports nearly half of all the plant and animal species in the world?

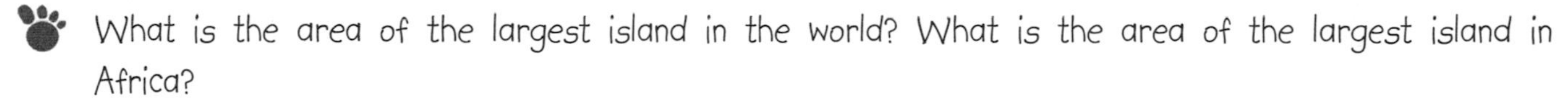
What is the area of the largest island in the world? What is the area of the largest island in Africa?

Day 3

What is the longest river in the world? Where are most of the world's rainforests located: near the prime meridian, near the equator, or along the Tropic of Cancer?

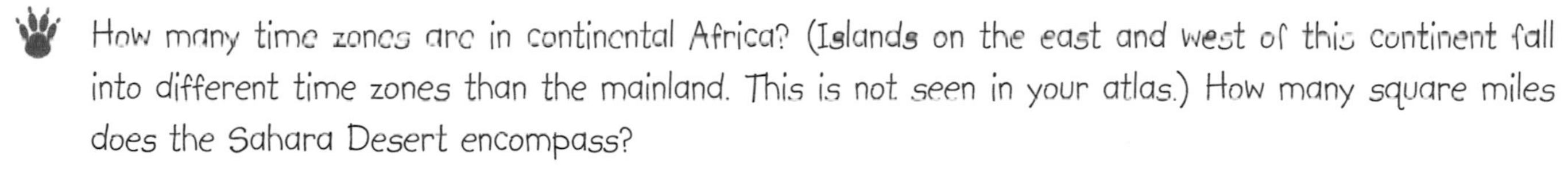
How many time zones are in continental Africa? (Islands on the east and west of this continent fall into different time zones than the mainland. This is not seen in your atlas.) How many square miles does the Sahara Desert encompass?

What is the deepest point in the Atlantic Ocean? What lake has the greatest volume of water, holding more than three times the amount of water than the second largest lake?

Day 4

In what continent is the world's largest rainforest? (hint: see world climate map) What two mountain ranges separate Europe from Asia? (hint: see world physical map)

What is the highest place in Europe? What two European countries are the smallest in the world?

Where is nearly 90% of the earth's ice located? What is the world's deepest lake?

POINTS OF INTEREST

Review

Use this week to review your work.

- Organize your notebooks and finish any incomplete projects.
- Review countries and capitals of North America, South America, Europe, and Africa.
- Use your flash cards and jigsaw puzzles.
- Work your crossword puzzles again.
- Do any art project you wanted to do but lacked the time.
- Look over your maps.
- Finish any incomplete thematic maps of each continent.

GEOGRAPHY TRAILS

Week 19 - Asia

Day 1

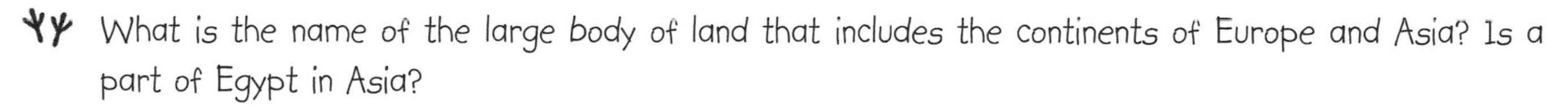

What is the name of the large body of land that includes the continents of Europe and Asia? Is a part of Egypt in Asia?

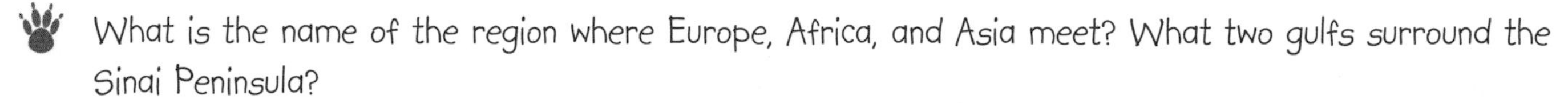

What is the name of the region where Europe, Africa, and Asia meet? What two gulfs surround the Sinai Peninsula?

What two seas does the Suez Canal link? What countries have coastline on the Persian Gulf?

Day 2

What seas surround the oil-rich peninsula that includes Saudi Arabia? Name four other countries in the Arabian Peninsula.

What two gulfs separate Iran and the Arabian Peninsula? What countries in the Arabian Peninsula are at or below the Tropic of Cancer?

Between what two rivers is Mesopotamia located? What countries share a border with Israel?

Day 3

Through what countries do the Euphrates and Tigris Rivers flow? Which country does not share a border with Israel: Jordan, Saudi Arabia, or Lebanon?

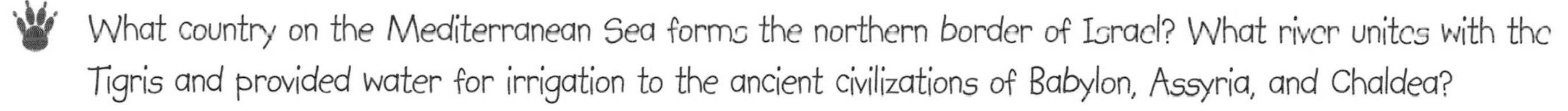

What country on the Mediterranean Sea forms the northern border of Israel? What river unites with the Tigris and provided water for irrigation to the ancient civilizations of Babylon, Assyria, and Chaldea?

What is the capital of the country located south of Turkey in the Mediterranean Sea? What Middle East country is located on the Gulf of Aden?

Day 4

Name four countries with a coastline on the eastern border of the Mediterranean Sea. What country is east of Iraq?

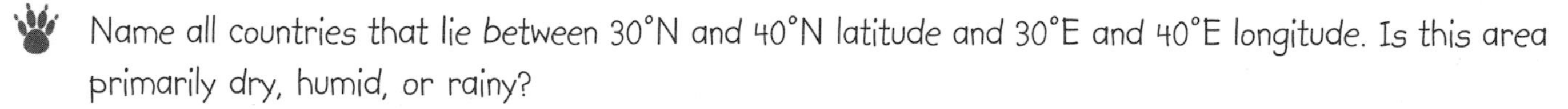

Name all countries that lie between 30°N and 40°N latitude and 30°E and 40°E longitude. Is this area primarily dry, humid, or rainy?

Is the Arabian Peninsula more plains, hills and low tablelands, or mountainous? What term best describes the climate of Iran: tropical, semiarid, or rainy?

POINTS OF INTEREST

Middle East

Mapping

Middle East (OMB 37)

- Label these bodies of water in blue: Mediterranean Sea, Gulf of Suez, Red Sea, Gulf of Aden, Arabian Sea, Indian Ocean, Gulf of Oman, and Persian Gulf.
- Draw and label the Tigris River, Euphrates River, Suez Canal, and Nile River.
- Label the Arabian Peninsula.
- Shade the Sinai Peninsula in purple and label it in the legend box.
- Place a triangle at Mount Ararat in Turkey and label.
- Shade map with browns, orange, yellow, and greens for land elevations, and include in the legend what elevation each color represents.
- Label each country listed below. Also label Afghanistan and Pakistan.

OR start on Asia "Map-It" (TUG 190) to complete in the next three weeks.

OR start on Asia Outline Activity (TUG 208-210) to complete in the next three weeks.

Trail Blazing

Read about the Middle East (TUG 182). Find travel videos of places in the Middle East.

The Middle East produces one-third of the world's oil. Look at an energy thematic map to determine which area of the Middle East has the highest energy reserves.

Study energy. Learn about fossil fuels and alternative sources of energy.

Cook a meal typical to the people of Israel from *Eat Your Way Around the World* or from another source.

The very dry climate of the Middle East has served as a perfect preservative for the history of this region. Study archaeology. Do your own archaeology experiment with the Chocolate Chip Cookie Dig. (TUG 80) Bonus: when you're done, you get to eat your project!

Learn the countries and capitals of this region. Make your own "Geography Concentration" game. Start with the following countries:

Turkey	Cyprus	Syria	Lebanon	Israel	Iran	Iraq	
Yemen	Egypt	Saudi Arabia	Kuwait	Bahrain	Oman	Qatar	United Arab Emirates

Learning the Five Themes of Geography - LOCATION

LOCATION, the first of five geography themes, is simply knowing where a place is. You have been getting practice in locating places in your atlas already. How well do you know location in your daily life? Practice this around your home and school. Where is the library: east of your home, or west? Where is the nearest park? Learn to use cardinal directions and continue reading maps. Make a map of your neighborhood. (Teachers: for more background on the five themes of geography, see TUG 43-52.)

Learn to use a compass (TUG 45) and begin to develop the life-long habit of knowing just where in the world you are!

Geography Notebook

Continue watching for and clipping newspaper articles about events occurring in the Middle East. There is no shortage of articles on the conflict and uprisings in this part of the world.

Jerusalem's Dome of the Rock is a popular and controversial landmark in Israel. Its construction began in A.D. 691. Find a picture of it and other landmarks to put in your notebook. Include a map identifying where they are located.

GEOGRAPHY TRAILS

Week 20 - Asia

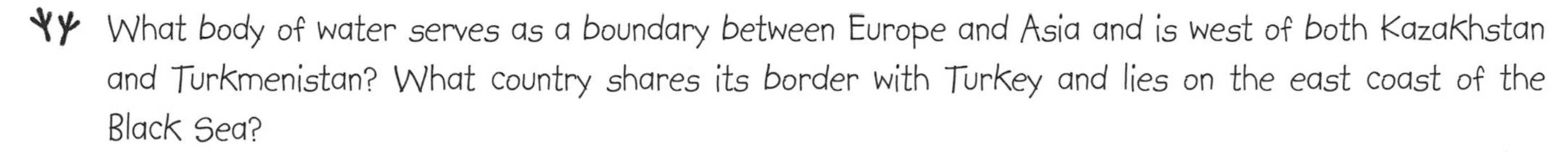

Day 1

- What body of water serves as a boundary between Europe and Asia and is west of both Kazakhstan and Turkmenistan? What country shares its border with Turkey and lies on the east coast of the Black Sea?
- What sea is located between Kazakhstan and Uzbekistan? What sea is west of Turkmenistan and Kazakhstan?
- What Asian countries have a coastline on the Caspian Sea? What mountains form the northern boundary of Georgia and pass through Azerbaijan?

Day 2

- What body of water is east of Azerbaijan? Which country is NOT next to China: Kyrgyzstan, Tajikistan, or Uzbekistan?
- What two countries are located along the Caucasus Mountains and have a part of their boundaries within the border of Europe? Is the average elevation in Kazakhstan above or below 1500 feet?
- What countries form the southern boundary of Kazakhstan? In what country is the Kirghiz Steppe?

Day 3

- If you were at 45°N, 60°E, would you be walking or swimming? In what hemisphere is Asia located?
- What country is located south of Kyrgyzstan? Is the land area of Kyrgyzstan and Tajikistan lowland, plains, or mountainous?
- What is the name of the high altitude region, mostly in Tajikistan, extending to parts of China, Kashmir, India, and Afghanistan? What is the elevation of the Caspian Sea?

Day 4

- Is Kyrgyzstan east or west of 90°E longitude? Is most of the land in Afghanistan lowlands, plains, or mountainous?
- The Amu Darya River winds nearly parallel between the border of Turkmenistan and what other country? What is the climate around the Aral Sea?
- What term best describes the land in Turkmenistan: cropland, forest, or sparse grass? What inland body of water is located between Kazakhstan and Uzbekistan?

POINTS OF INTEREST

Central Asia: Former Soviet States

Mapping

Asia: former Soviet States (appendix 108)
Use the map provided in the appendix of this book.
- Shade in blue and label the Black Sea, Caspian Sea, and Aral Sea.
- Draw a zigzag line along the Ural Mountains and the Caucasus Mountains.
- Label the following countries and their capitals: Georgia, Armenia, Azerbaijan, Turkmenistan, Uzbekistan, Tajikistan, Kyrgyzstan, Kazakhstan.

Trail Blazing

Read about Asia (TUG 181, GTA, student atlas), find travel videos of places in Asia from the library, and and learn about the people of Asia.

The world's highest dam is Rogun Dam on the Vakhsh River in Tajikistan. It is 1066 feet tall. If you have not studied dams yet, do so. Read about the environmental impact of building dams.

Learn the countries and capitals listed above. Make more cards and add them to your Geography Concentration game. Shuffle and play with the countries from this week and last. Plan to add more next week.

Learning the Five Themes of Geography - PLACE
This theme may be better named "characteristics." It's as simple to understand as answering the question, "What makes this place unique?" Look for characteristics in a physical way such as climate, fauna and flora, land elevation, natural resources or in a human way such as architecture, culture, population density, language, economics, and so on. You have seen some of the various characteristics of the PLACE for yourself with the thematic maps and illustrated dictionary you have made.

Begin studying the weather (TUG 112-113). Learn how to identify clouds and predict the weather where you live. Pay attention to the climate maps of any area of study, and you will gain a better understanding of what it's like there. It helps to compare the places of study with your own place of residence.

Geography Notebook
Make your own treasure map that requires the use of a compass (TUG 82-83).

Make a crossword puzzle with countries and capitals from the last two weeks.

Using an almanac, chart statistics on Russia and these eight former Soviet Republics. Include area, per capita GNP (or GDP), infant mortality rate, population density, type of government. Analyze the data on your chart. How are you at graphing? Try graphing these figures for a more visual perspective (TUG 122). Do you think the citizens of these countries have a better way of life since they broke away from the communist U.S.S.R.?

Select 1-3 countries from this continent. Complete the Countries of the World: A Fact Sheet (GTA 171) for each country. Or follow the instructions from Week 11 to design your own sheet.

OR Select a country from this continent. Write a short essay from the choices in TUG 218. Feel free to include a sketch of the country's flag.

Do Asia challenge questions (TUG 211).

GEOGRAPHY TRAILS

Week 21 - Asia

Day 1

- What is the capital of Afghanistan? What river runs through Pakistan?
- Where is the Deccan Plateau? The tallest mountain in the world is in the Himalayas; name it.

- What is the capital of the country through which the Indus River flows? Bombay, India, is on the coast of what body of water?

Day 2

- What country is a peninsula in the Indian Ocean and the Arabian Sea? What country is an island in the Indian Ocean?
- What country is south of Afghanistan and west of India? In what ocean is Sri Lanka?

- What are the two mountain ranges in southern India? What large, high, flat land area lies between these two mountain ranges?

Day 3

- Is the Bay of Bengal in the Indian Ocean or the Pacific Ocean? What country is located both east and west of Bangladesh?
- Kathmandu is the capital of what country which lies high in the Himalayas? Which country has more of its economy from manufacturing and commerce: Bangladesh or Bhutan?
- What small country in the Himalayas lies on the 90°E line of longitude? What island nation southwest of India is the smallest country in Asia?

Day 4

- What mountains form a boundary for India, Nepal, Bhutan, and China and are the highest in the world? Through what two countries does the Ganges River flow?
- What creates a natural barrier that separates the two most populous countries in Asia? Which best describes Afghanistan: humid plains or dry and mountainous?
- In what two countries is the Ganges River located when it flows into the Bay of Bengal? What is the capital of the island country once known as Ceylon?

POINTS OF INTEREST

South Asia

Mapping

India (OMB 29)
- Label the Arabian Sea, Bay of Bengal, and Indian Ocean in blue.
- Trace the Ganges and Indus Rivers in blue and label.
- Put a triangle on Mount Everest and label.
- Shade the Himalayas and plateau of Tibet lightly with brown.
- Shade the Western Ghats and Eastern Ghats with orange, and label.
- Shade the Deccan Plateau yellow, and label.
- Label the Great Indian Desert.
- Label the following countries and their capitals: Afghanistan, Pakistan, India, Nepal, Bhutan, and Bangladesh.

Note: For consistency in map labeling, use uppercase and lowercase letters for city names and other places; use all CAPS when printing continent or country names. Use a dot within a circle for capital cities.

Trail Blazing

Read about south Asia, often referred to as the "Indian subcontinent" (TUG 183). Find travel videos from the library of places in this region.

Learn the countries and capitals listed above. Make more cards and add them to your Geography Concentration game. Shuffle and play with the countries from the past three weeks. Is it getting harder to play? Plan to add more next week.

Cook a meal typical to the people of India from *Eat Your Way Around the World* or from another source.

The wettest place in Asia is Cherrapunji, India, where 450 inches of rain falls each year! Study hurricanes, tornadoes, and thunderstorms. What factors contribute to the formation of storms? Explain each of these three storms. Learn about cold fronts and warm fronts. Describe the common weather instruments used for measurements.

Learn about the great Himalayas, the highest mountain range in the world. Make a salt dough or clay physical relief map of Asia. See how well you can form the Plateau of Tibet and the Himalayas. Only include the major islands.

Geography Notebook

Start a chart of Asia facts. Follow instructions from Week 6. Select ten countries from Week 19 and 21. You will add to this chart the next two weeks as you continue to study Asia.

India is a highly populated country. Look at a population density map. There are about one billion people living in India. Study life in this crowded country. Is there enough food? What is the quality of life? Write a report and place in your notebook. Include information about and a picture of the Taj Mahal.

Learn about animals of Asia and add to the animal notebook or section in the geography notebook.

Add more Countries of the World Fact Sheet pages (GTA 171) to your notebook.

Geography Through Art

- Learn about Mughan Indian miniature painting.
- QuickSketch: Elephant
- Draw your own Taj Mahal.
- Learn the art of batik.

GEOGRAPHY TRAILS

Week 22 - Asia

Day 1

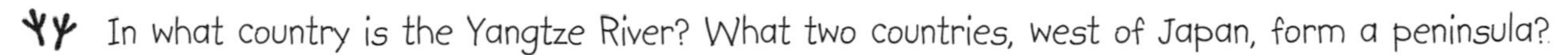

In what country is the Yangtze River? What two countries, west of Japan, form a peninsula?

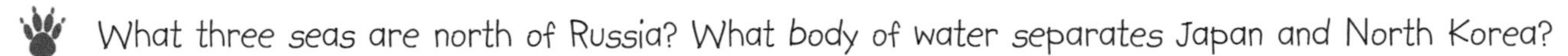

What three seas are north of Russia? What body of water separates Japan and North Korea?

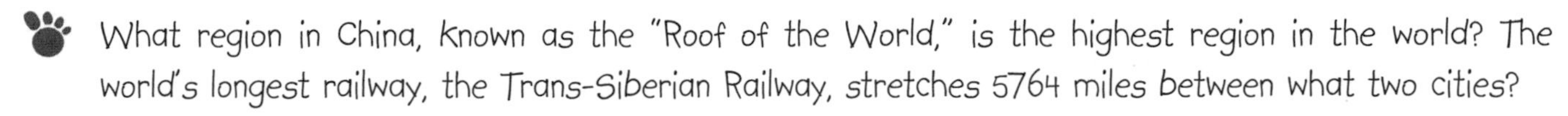

What region in China, known as the "Roof of the World," is the highest region in the world? The world's longest railway, the Trans-Siberian Railway, stretches 5764 miles between what two cities?

Day 2

What is the world's largest country? Name two rivers in Russia whose water is eventually carried into the Arctic Ocean.

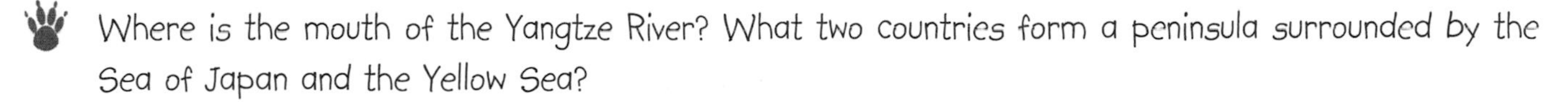

Where is the mouth of the Yangtze River? What two countries form a peninsula surrounded by the Sea of Japan and the Yellow Sea?

What country is a series of islands that stretch out for approximately 1900 miles and has Asia's most powerful economy? What is the dry region in northern China and southern Mongolia?

Day 3

What country is surrounded by Russia and China? What is the capital of China?

Where is the Plateau of Tibet? What island nation in east Asia is 75% mountainous where numerous natural hazards such as tsunamis, earthquakes, and volcanic eruptions are commonplace?

What is the most populous country in Asia? What sea is off the eastern coast of Russia?

Day 4

What country north of the Tropic of Cancer is a chain of volcanic islands in the Pacific Ocean? What island nation is located on the Tropic of Cancer?

What passageway of water connects the East China Sea and the South China Sea? What desert is located between Mongolia and China?

Which sea is deeper: the Sea of Japan or the East China Sea? What is the capital of the island country located at the Tropic of Cancer?

POINTS OF INTEREST

Russia, Mongolia, and the Far East

Mapping

China and Mongolia (OMB 23)

- Draw and label the Yangtze and Huang Rivers in blue.
- Label Sea of Japan, Yellow Sea, East China Sea, and South China Sea in blue.
- Shade the Gobi desert in orange and label.
- Label China, Mongolia, North Korea, and South Korea and their capitals.
- Shade Russia gray.

Asia (OMB 13)

- Label Russia, Mongolia, China, Japan, India, and Taiwan.
- Label Sea of Okhotsk, Sea of Japan, Yellow Sea, East China Sea, South China Sea, Philippine Sea, Bay of Bengal, Arabian Sea, Red Sea, Indian Ocean, and Pacific Ocean in blue.
- Write a RR at both Moscow and Nakhodka in Russia. These are the starting and ending points of the longest railway in the world.
- Label Malaysia and Indonesia (all CAPS).

Trail Blazing

Read about China, Korea, and Japan (TUG 185-187).

Continue learning the countries and capitals of Asia. Make more cards with the countries listed above and add them to your Geography Concentration game.

Cook a meal typical to the people of China, Japan, or South Korea from *Eat Your Way Around the World* or from another source.

Learning the Five Themes of Geography - RELATIONSHIP

This theme focuses on how people relate to their environment. Consider why people live where they do and how the climate, plants, and animals affect quality of life. For example, the climate and land in Asia are perfectly suited to rice farming. Over 90% of the world's rice is produced in this continent. How has this affected the culture and life?

Look at several population density maps. What areas are most heavily populated: mountainous regions, deserts, rainforests, or arctic regions? Compare the population density with all other thematic maps available. Look at manufacturing, land use, and environments. See if you can draw your own conclusions about what influences the choices people make of where to live.

Gather pictures of houses from around the world. What makes these homes different? What factors influence the architecture, style, and materials used to build these homes?

The Great Wall of China is the only man-made structure that can be seen from space. Study this marvel, its many construction stages, and its history. Place this study and pictures in your Geography Notebook.

Geography Through Art

There are many interesting art project choices this week from China. Projects from Japan are given next week.

- QuickSketch: Hall of Prayer for Good Harvest
- Chinese paper cutting
- QuickSketch: Great Wall
- QuickSketch: Pagoda
- Chinese scrolls
- Chinese calligraphy
- Terra Cotta soldiers
- QuickSketch: Dragon
- Tangram Puzzle

GEOGRAPHY TRAILS

Week 23 - Asia

Day 1

- Is Laos north or south of Cambodia? What country is a large archipelago between the Indian Ocean and the Pacific Ocean?
- What large area of water lies within the coastline of Thailand, Cambodia, and Vietnam? What nation is made up of thousands of islands that span 95°E to 141°E longitude?
- What countries make up Indochina? Saigon, Vietnam, renamed for the founder of the Indochina Communist Party in 1975, is now called what?

Day 2

- What country is west of Thailand? What country, east of Vietnam, is a chain of islands in the Pacific Ocean?
- Between what two countries does the Strait of Malacca flow? What is the capital of the Philippines?
- What is the world's most populous island (over 118 million people!)? What country is an archipelago north and east of the Sulu Sea?

Day 3

- Bangkok is the capital of what country? What Asian country near the equator has two distinctively separated landmasses? (One is part of an island and one is part of the Malay Peninsula)
- Is Malaysia located east or west of 105°E longitude? Which country has a higher population density: Myanmar (Burma) or Brunei?
- What three countries are located on the Malay Peninsula? The Mekong River forms part of the western border of what country?

Day 4

- What capital city is located on the southern tip of the Malay Peninsula? (hint: see western part of Malaysia) The Ayeyarwady River flows into the Bay of Bengal through what country?
- Which city has a more urban environment: Jakarta, Indonesia or Hanoi, Vietnam? The Sulu Sea helps form the western boundary of what country?
- What river flows through the middle of the country located west of Thailand? Which city does NOT have a tropical climate: Jakarta, Indonesia; Hanoi, Vietnam; or Begawan, Brunei?

POINTS OF INTEREST

Southeast Asia

Mapping

Indonesia (OMB 30)

- Mark the equator with a dotted line. Label with 0° in the left and right margin (border) of the map. Write "Equator" down the margin.
- Label the following bodies of water in blue: South China Sea, Philippine Sea, Indian Ocean, Sulu Sea, Strait of Malacca, and Gulf of Thailand.
- Shade Thailand pink, Philippines red, Malaysia green, Brunei orange, and Indonesian islands purple. Circle Singapore. Put these in a legend on the upper right corner of the map.
- Label the northern tip of Australia.
- Label Borneo (which is a part of Indonesia).

Vietnam (OMB 48)

- Label the following countries and their capitals: Myanmar, Thailand, Laos, Cambodia, and Vietnam.
- Draw Ayeyarwady (Irrawaddy) and Mekong Rivers and label in blue.
- Label Hanoi and Ho Chi Minh City in Vietnam.

Label the Gulf of Thailand and Adaman Sea in blue.

Trail Blazing

Continue learning the countries and capitals of Asia. Make more Geography Concentration cards with the countries listed below and play the game.

Brunei	Cambodia	Myanmar	Laos	Vietnam
Thailand	Philippines	Malaysia	Singapore	Indonesia

Some gems found in Asia include diamond, emerald, ruby, jade, zircon, sapphire, and moonstone. Find where these are mined and shade the regions on a separate Asia map. Research one or two gems to learn more about how they are mined, how they are used, and the effect on the economy.

Learn about the principal crops of southeast Asia. How does the climate and soil contribute to agriculture?

On December 26, 2004 a powerful undersea earthquake off the coast of Sumatra triggered the largest tsunami in four decades, leaving over 288,000 dead or missing victims. Learn what countries were affected and mark them on a map. Study how undersea earthquakes create tsunamis.

Geography Notebook

Learn about the orangutans in Borneo. Learn about other animals of Asia and add to the animal notebook or the animal section in the geography notebook.

Add ten new countries to the chart of Asia facts.

Answer challenge questions for Asia (TUG 211).

Make your own set of thematic maps of Asia. Follow the instructions from Week 3.

Geography Through Art

Choose from these interesting projects from the Far East and Southeast Asia.

- Fish printing
- Divider Screens
- Folded Fan
- Sumi-e ink painting
- Girl's Day Dolls
- Make your own Hmong story cloth.
- Origami
- Boy's Day Fish Banners

GEOGRAPHY TRAILS

Week 24 - Australia and Oceania

Day 1

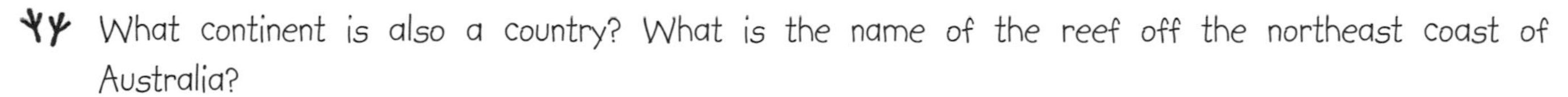

What continent is also a country? What is the name of the reef off the northeast coast of Australia?

Is New Zealand an island of the country of Australia, or a country in Oceania? The Great Barrier reef lies between Australia and what sea?

What land areas comprise Oceania? What are the three deserts located in the western half of Australia?

Day 2

What Tropic passes through the midst of Australia? What country is made up of the North Island and the South Island and is located east of the Tasman Sea?

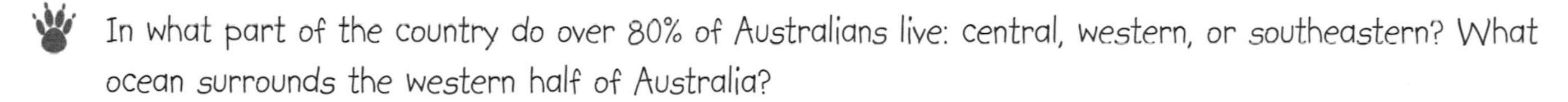

In what part of the country do over 80% of Australians live: central, western, or southeastern? What ocean surrounds the western half of Australia?

Where is the Murray-Darling system, the main river in Australia, located? Is the average elevation of New Zealand above or below 1000 feet?

Day 3

What is the largest city in Australia? Australia is the world's largest producers of which: sheep and wool, beef and milk, or rice and corn?

What are native Australians called? What strait passes between the North Island and the South Island of New Zealand?

What is the longest coral reef in the world? What is the nickname for Australia's mostly uninhabited interior?

Day 4

What Australian island is located south of Melbourne? Is most of the land in Australia and New Zealand used for farming, ranching, or manufacturing?

What river flows through the northern part of New South Wales Territory? What type of climate describes the region east of the Great Dividing Range in Australia and most of New Zealand?

What best describes the climate of New Zealand: hot and rainy, very dry, mild and rainy, or tropical? What is the capital of the largest country in Oceania?

POINTS OF INTEREST

Australia and New Zealand

Mapping

Australia and New Zealand (OMB 14)

- Label the following bodies of water in blue: Indian Ocean, Timor Sea, Torres Strait, Coral Sea, Pacific Ocean, Gulf of Carpentaria, Cook Strait, and Lake Eyre.
- Draw and label the Murray/Darling Rivers.
- Mark the Tropic of Capricorn at 23.5° S with a dot-dash line and label it.
- Label these countries and their capitals: Australia, New Zealand, and Papua New Guinea.
- Label the following territories and states: Western Australia, South Australia, Northern Territory, Queensland, New South Wales, Victoria, and Tasmania.
- Label the North Island and South Island of New Zealand and the Solomon Islands of Melanesia.

OR do Australia "Map-It" (TUG 195).

OR do Australia Outline Activity (TUG 215-216).

Trail Blazing

Read about Australia (GTA, TUG 191-194, student atlas) and find travel videos of places in Australia and Oceania. (Australasia is sometimes used to refer to Australia, New Zealand, and Oceania.)

Make a jigsaw map of Australia and New Zealand. Follow instructions from Week 7. See how fast you can put the puzzle together.

Australia is the only country that is also a continent. The country of Australia includes the island of Tasmania. The South Pacific region around Australia that includes New Zealand, Papua New Guinea, and the island groups known as Melanesia, Micronesia, and Polynesia, has come to be known as Oceania. Even though maps of Australia include New Zealand, it is important to know that New Zealand is its own independent country. Learn about New Zealand.

Learn about the Sydney Opera House.

Learning the Five Themes of Geography - MOVEMENT

We interact with the world around us on a daily basis without even realizing it. Start reading the "made in" labels on clothing, electronics, and toys, and you will get the idea. How did all this stuff get from all of these different countries in the world to your home? Goods can travel by boat, train, truck, or plane. Information through history has moved by courier, Pony Express, telegraph, newspaper, telephone, TV, and Internet - the faster the better! Watch how the movement of goods relates to the climate, culture, and resources available.

Geography Notebook

Australia has a number of animals indigenous to nowhere else on earth. Study the animals of this country. Notice the climate and food needed by these interesting animals and see if there is any other place on earth where they could survive. Add to your animal notebook or start a new notebook (or a new section in your geography notebook) about the uniqueness of Australia.

Answer challenge questions for Australia and Oceania (TUG 216).

Geography Through Art

- Make a boomerang.
- Aboriginal Stone Painting
- QuickSketch: Kangaroo
- Aboriginal Dot Painting
- Aboriginal Bark Painting
- QuickSketch: Sydney Opera House

GEOGRAPHY TRAILS

Week 25 - Australia and Oceania

Day 1

- Is Australia in the Northern Hemisphere or Southern Hemisphere? Is Papua New Guinea north or south of Australia?
- Oceania is made up of Australia, New Zealand, Papua New Guinea and nearly 25,000 islands in the South Pacific; name one of the South Pacific island groups? Are the Marshall Islands in Oceania?
- What are the three main island groups of Oceania? What islands in Melanesia are located near the coordinates of 8°S and 160°E?

Day 2

- Is Fiji in Oceania? What is the capital of Australia?
- Which region is in the Western Hemisphere: Polynesia, Micronesia, or Melanesia? What Asian country shares its eastern border with Papua New Guinea?
- What is the second largest island in the world? In what two continents is it located?

Day 3

- What is the highest point in the country of Australia? What part of Oceania has rainforests: southern, central, or northern?
- What tropic parallel transverses Australia? Is New Zealand east or west of Tasmania?
- What is the elevation of the highest point in Oceania, and where is it? What U.S. state is home to the wettest place in Oceania?

Day 4

- Places with the warmest climates are generally located between what two lines of latitude? Australia has more desert regions than any place in the world except the Sahara; name one desert located in the western half of Australia.
- What type of climate do Papua New Guinea, Melanesia, Micronesia, and Polynesia have in common? What is the largest group of coral reefs and islands in the world?
- Oceania's third largest country has over 50 million sheep; what is its capital? What is the lowest elevation in Oceania?

POINTS OF INTEREST

Oceania

Mapping

Pacific Rim (OMB 40)

This may be a challenging exercise, but you can do it! Use the Pacific Rim, or Oceania Political map in your atlas as a guide. Do you notice a portion of the Western Hemisphere on the right of the map?

- Label Australia, New Guinea, New Zealand, Indian Ocean, and Pacific Ocean.
- Mark the international date line with a dotted line and label at the bottom of the map.
- Circle the Hawaiian Islands in red.
- Locate and draw a box around the island groups of Melanesia, Micronesia, and Polynesia. (It's not necessary to identify each South Pacific island perfectly when marking the island groups.)
- Label the names of your choice of islands in each group.

Trail Blazing

Read about New Zealand and Papua New Guinea and do "Map-It" (TUG 195) on a separate map of Australia.

The Great Barrier Reef covers 80,000 square miles and is the largest group of coral reefs and islands in the world. It is made up of over 2500 separate coral reefs with over 400 types of coral and is home to around 1500 species of fish. Study this fantastic physical feature of Oceania and obtain pictures of several kinds of fish and coral found there. Add this to the Australia notebook begun last week.

Learn about the peoples of Australasia and Oceania.

Cook a meal typical to the people of Australia or New Zealand from *Eat Your Way Around the World* or from another source.

Learning the Five Themes of Geography - REGIONS

The earth can be divided into any number of regions. Anything that is common can be deemed a region. You've been mapping regions every time you make a thematic map or shade a desert. Most of these regions have been of a physical nature. Human regions include common language, politics, religion, economics, and…? What human regions can you name?

Geography Notebook

Make your own set of thematic maps of Australia and Oceania. Follow instructions given in Week 3.

Make a chart of Oceania facts. Place country names down the left side of the chart. Add columns across the top and choose labels from these topics: capital, area, currency, language, principal religion, and natural resources. See how many cells in the chart you can fill in from your resources, the Internet, or an almanac. Use these countries:

Australia	New Zealand	Papua New Guinea	Fiji Islands	
Solomon Islands	Samoa	Tuvalu	Micronesia	Marshall Islands

Select 1-3 countries from this continent. Complete the Countries of the World: A Fact Sheet (GTA 171) for each country.

OR Select a country from this continent. Write a short essay from the choices in TUG 218. Feel free to include a sketch of the country's flag.

Geography Through Art

- Make a didgeridoo.
- Make a spiny anteater with clay.

GEOGRAPHY TRAILS

Week 26 - Antarctica

Day 1

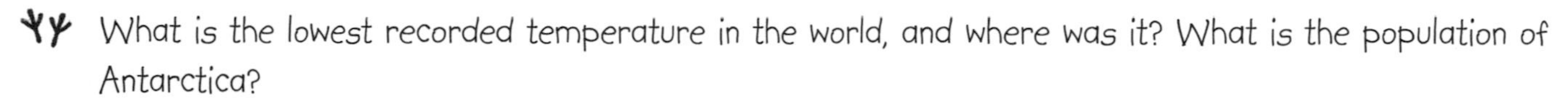

What is the lowest recorded temperature in the world, and where was it? What is the population of Antarctica?

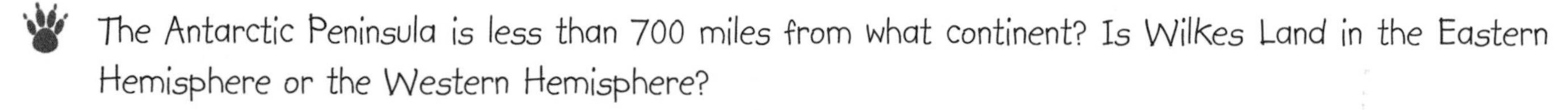

The Antarctic Peninsula is less than 700 miles from what continent? Is Wilkes Land in the Eastern Hemisphere or the Western Hemisphere?

What sea and ice shelf are south of New Zealand? Is the American Highland in the Eastern Hemisphere or the Western Hemisphere?

Day 2

What ocean surrounds Antarctica? What covers most of the land of Antarctica?

What kind of climate does Antarctica have? What is the mountain range that spreads from Victoria Land to Rockefeller Plateau?

Where is the Antarctic Peninsula? What is the name of the basin located at the South Pole?

Day 3

What is the name for the southernmost part of the earth located near the center of Antarctica? What sea is along the Ross Ice Shelf?

Is the South Magnetic Pole located on the continent of Antarctica? What ice shelf is located near Lambert Glacier?

Is Enderby Land nearer to Australia, Africa, or South America? Where is the Magnetic South Pole located?

Day 4

What is the highest point in Antarctica? Which ice shelf is farthest north: Ross Ice Shelf or Ronne Ice Shelf?

What cape is located in the Western Hemisphere near 10°W latitude? The highest point in Antarctica is located on what mountains?

What four seas form boundaries with Antarctica? Along which ice shelf do the most mountain peaks rise?

POINTS OF INTEREST

Antarctica

Mapping

Antarctica (OMB 11)
- Label the following lands: Wilkes, Enderby, Queen Maud, Ellsworth, Marie Byrd, Graham, and 🐾🐾 Coats.
- Shade the Transantarctic Mountains and label.
- Label the South Magnetic Pole.
- Label these ice shelves: Amery, Filcher, Ronne, and Larsen.
- Label these bodies of water in blue: Weddell Sea, Ross Sea.
- Label these oceans: Indian Ocean, Pacific Ocean, and Atlantic Ocean, (or label the Southern Ocean).

Trail Blazing

There are seven countries that claim land on Antarctica. None are recognized by the Antarctic Treaty of 1961 that mandates the continent be used only for peaceful purposes and scientific study. Currently over 44 research stations operated by 23 different countries are established there. What are all those research scientists studying? For great information regarding the research stations log on to: http://www.nsf.gov/od/opp/antarct/start.htm

Nearly 98% of the land is under ice sheets, three miles thick in some places. Some believe, that if all the ice in Antarctica melted it would raise the level of the oceans by over 200 feet, flooding coastal cities all over the world. Study glaciers and glaciology. Write a report or do an oral report on your findings.

Do some of your own research about Antarctica. What is the climate like? In what part of the continent is the climate mildest? Even though there are valuable resources under all that ice - including iron ore, copper, lead, zinc, gold, silver, and oil - mining is prohibited.

Learn about Norwegian Roald Amundsen's expedition to the South Pole. How long did it take? What supplies did he have? How does this compare to supplies available today?

Four species of penguins live on Antarctica. Read Penguins at Home: Gentoos of Antarctica or Antarctica (both books written by Bruce McMillan) to learn more about them.

Use the geography terms flash cards to review the many different geography terms you have learned (TUG 263-280). Add any new terms to your Illustrated Geography Dictionary.

This assignment is repeated from Week 2. If you have not already done so, study the Southern Ocean. The International Hydrographic Organization (IHO) declared and demarcated a fifth ocean and named it the Southern Ocean. It includes all the water below 60°S to the boundary of Antarctica. Learn more about the fifth ocean and how it came to be established. Try using about.com or the CIA Factbook online at: http://www.cia.gov/cia/publications/factbook/geos/oo.html

Geography Notebook

Draw the boundaries of each of the seven land claims on an outline map of Antarctica.

Study water and learn about the water cycle. Use an encyclopedia, library books, or science textbooks. Learn the three states of water and about water erosion. Draw a diagram of the water cycle.

Study other animals that make their homes on and around Antarctica. Here are a few to get you started: albatross, sea lion, whales, and krill. Add to your animal notebook.

GEOGRAPHY TRAILS

Week 27 - Review

Day 1

- What is the largest continent in the world? What is the world's largest island? (hint: located at 75°N)

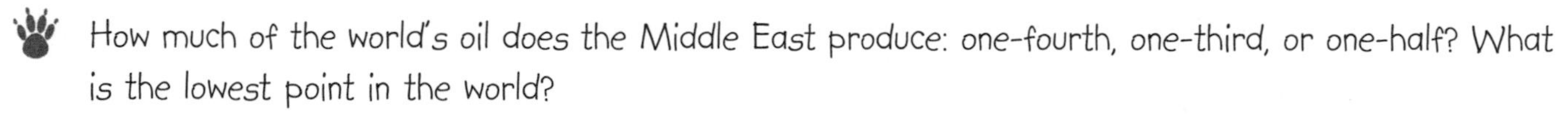

- How much of the world's oil does the Middle East produce: one-fourth, one-third, or one-half? What is the lowest point in the world?
- What country has the largest petroleum reserve? How tall is the highest point in North America?

Day 2

- What is the climate of Indonesia? What is the climate of central Australia?

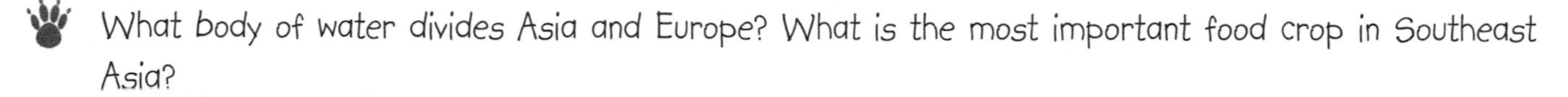

- What body of water divides Asia and Europe? What is the most important food crop in Southeast Asia?
- Name three countries in Asia that have time zones that vary by half an hour? What is the greatest known depth of the ocean?

Day 3

- What is a country? Which ocean is the largest?
- Why does Australia have such a large population of marsupials? Australia and New Zealand produce nearly half of the world's supply of: wheat, wool, or beef?
- What is the equatorial speed of the earth's rotation? What is the speed of the earth's orbit around the sun?

Day 4

- What is the highest mountain in the world? What is the most populated continent?
- What science station in Antarctica has the largest number of people on staff? What passage lies between South America and Antarctica?
- What is the most populous city in the world? What is the most populous country?

POINTS OF INTEREST

The World in Review

Mapping

World (OMB 49)
- Make four thematic maps showing world regions. Use a separate world map for each theme. Include a leg end explaining the meaning of each color used.
  - Maps 1 and 2 - Natural characteristics: climate and physical
  - Maps 3 and 4 - Human characteristics: economics and population density
- Time zones - Make your own time zone map using color columns to depict each zone. (You should be able to find a time zone map in your atlas.)
  1. Mark the international date line and the prime meridian in red.
  2. Place each hour of time across both the top and bottom of the map.
  3. Color code each time zone column in top and bottom margins and on the landmasses they represent.

Trail Blazing

The world is divided into 24 standard time zones. Some places, such as Iran, Tibet, and India, do not have standard one-hour changes. Learn how to read a time zone map (TUG 27) and practice with one you have made or with one in your atlas or encyclopedia. Determine the following:

1. When it is 3:00 P.M. in your time zone, what time is it in London? Nairobi? Beijing? Lima? Rio de Janeiro? Sydney?
2. When it is 11:00 A.M. in your time zone, what time is it at those same locations?

Learn about the solar system. Where is the earth in relation to the other planets in our system? Draw a diagram depicting the planets in order, showing the diameter of each planet and how far each is from the sun. What is the Milky Way?

Learn about the seasons and how the position of the earth's tilt relates to weather and changing seasons? Draw a diagram of the seasons using the earth and its position with the sun. Do you understand why the seasons in the Southern Hemisphere are opposite from Northern Hemisphere?

Learn about the moon and how it affects the changing tides. Learn about the phases of the moon and terms such as "waxing" and "waning" associated with them. Observe the night sky and draw the phases of the moon for the next 2-4 weeks.

Learn about eclipses. What is a lunar eclipse? What is a solar eclipse? Draw a diagram of each.

Geography Through Art
Choose from any additional projects included in this book that you have not previously completed.

Around the World in Eighty Days

A Unit Study

by Sylvia Hemme and Cindy Wiggers

Objective #1: To study a "classic" Jules Verne novel in an appealing manner which will propel the students to read other works by Jules Verne on their own.

Objective #2: To make the learning of geography, spelling, vocabulary, science, and social studies fun and interesting.

Introduction: Assign the students to do a one- to three-page research paper on the life of Jules Verne. This will help students understand the life of Verne, what interested him, and what knowledge he gained in order to write his books.

Life of Jules Verne

I. Childhood

A. Born February 8, 1828 in Nantes, France
B. Eldest of five children
C. Read classic children's literature
D. Entered boarding school at age nine
E. Preferred daydreaming, drawing, and writing to studying

II. Post-School Days

A. Worked in the theater for over five years, writing on the side
B. Quit theater to pursue serious writing
C. Married a widow with two little girls when he was twenty-nine years old
D. To support family, wrote early in the morning and worked during the day

III. Success

A. At age thirty-five, successfully wrote science fiction books for young readers
B. Success continued into adult science fiction
C. Became a successful, wealthy, and famous writer
D. Wrote constantly until the day he died, at age seventy-seven

Some Other Books by Jules Verne

Five Weeks in a Balloon
Twenty Thousand Leagues Under the Sea
A Journey to the Center of the Earth
From the Earth to the Moon
Around the Moon

Instructions

A word of caution: Disney's *Around the World in Eighty Days* movie is only loosely based upon the novel. Some publishers have followed up with a revision of Verne's classic novel, adapted to coincide with Disney's film. For this unit study be sure you use the original book and not this newer version.

Scheduling

The *Around the World in Eighty Days* unit study is set up for nine weeks (starting with Week 28 of this *Trail Guide*); however, you may need to adjust the assignments to meet your school plan. To complete the study in nine weeks you will read four chapters each week and five chapters the last week. As with any unit study guide, there are more assignment choices than possible to accomplish. Choose those that meet your objectives, suit the student's learning style, and instill a delight in learning.

Oral Reading

Read this book out loud. Take turns reading aloud, no matter how old or young the children. For the younger grades, you may choose to read the condensed version. Since the chapters are short, try to read the "real" book to the students, and explain anything they do not understand. This will stretch students mentally in a positive way. Perhaps they then will be interested in reading the condensed version on their own. A list of related books for younger children is included in the appendix.

Discussion

Questions for discussion are provided for each chapter. Use them to assess comprehension, as a springboard for further topics of discussion (the character qualities of each of the three main characters, for example), and to keep track of the progress of the journey. Answers to the questions are generally not provided as they are easily obtained from reading the chapter.

Assignments and Activities

Mapping, projects, and other assignments are listed for each week by chapter. If you are reading four chapters a week you won't have time to do all the activities, so don't even try. Select from the list those which your students will most enjoy.

Mapping/Geography

- Use the following outline maps and a student atlas to perform the map work. Students should add to the same maps throughout the study and not start with a fresh map each time a map is needed.

British Isles	Europe
Mediterranean Sea	Middle East
Asia	United States
World	

- Check out various library books and videos on the topics and locations listed to see the places and learn more of the various cultures from around the world.
- Add any of these fun geography resources to aid in learning the geography of this book:

Geo-Safari	Where in the World Is Carmen San Diego?
World Wise Games	RealEarth GlobeMap

Instructions

Arts and Crafts

You will be reminded each week to choose an art or craft project. Depending upon your schedule, assign one project every 1-2 weeks. Choose from the list below, adapt these to suit your needs, or develop ideas of your own.

- Choose one or more chapters to illustrate by creating a miniature 3-D exhibit.
- Draw or paint a scene from the chapter.
- Make travel brochures of any of the places in the text.
- Draw pictures of the main characters using the descriptions in the reading.
- Choose from projects in *Geography Through Art* appropriate to the country where the story takes you.

Additional Research

You may wish to select from a variety of optional ideas to expand the chapter. Assign research papers, additional books to read, oral reports, or other assignments for added learning experience. Topics for further study are listed for nearly every chapter. Students use encyclopedias, magazines, library books, or the Internet. Select only one or two each week, not one for each chapter.

Spelling and Vocabulary

A list of words from the chapter is given for the students to look up in the dictionary. Be sure students select the definition used in the context of the passage in the book. Arrange various other spelling exercises such as dividing into syllables, alphabetical order, using them in original sentences, scrambling, puzzle searches, etc., and test the meaning of words. Younger children can learn to use the dictionary in looking up the words; however, you shouldn't expect them to memorize the meaning or learn how to spell the words.

Math

Younger students can learn to tell time in a fun way with this book. Buy a clock stamp and when any time is noted, have the child stamp the clock and draw in the hands. Teach the way to write out the times and the different ways to say them (example: 6:45 or a quarter to 7:00 or fifteen minutes until seven).

Calculating the time: older students can figure the time according to Passepartout's watch, which he kept set on London (Greenwich Mean) time. This is a great opportunity to use a time zone map.

Calculating exchange rates: find current exchange rates in the newspaper or Internet. Calculate the value of British pounds compared to American dollars.

Calculating average rate of travel: calculate the approximate miles traveled each leg of the journey and divide by the number of hours it took to travel to determine the average miles per hour traveled.

Writing Projects

There are a variety of opportunities for students to develop and practice writing skills. Besides research assignments, here are three more options:

1. Write a five-page book report.
    - Include at least one illustration from the book.
    - Include one character study (detailed description of one character in the book -- not only outward appearances but also inward thinking).
    - Edit and re-write for final copy.
2. Make a journal as though you were traveling with Phileas Fogg and his companion. Write daily journal entries, including where the main characters are each day and what form of transportation they use.
3. Write newspaper articles informing the public of Phileas Fogg's progress and the impending arrest intended to end his trip prematurely.
4. Place all completed writing projects in the Geography Notebook.

Five Themes of Geography

Keep in mind the five themes of geography while reading this novel. These were introduced in weeks 19, 20, 22, 24, and 25 respectively.

1. Location
2. Place
3. Relationship
4. Movement
5. Regions

Movement and location are particularly evident as students observe the different forms of transportation used by Phileas Fogg on his journey around the world. Why was each form chosen, and what advantages or disadvantages did it present? Keep a travel record of what form of transportation was used, from what location to what location, how many miles between each leg of the journey, and how long it took. You can usually get the miles and time from the text. Discuss how long those parts of the journey would take today by boat, train, automobile, or plane.

Finale

To wrap up this unit, it is fun to watch a video of the movie after you have finished reading the book. You may want to show students various versions of this story. Instruct them to watch with an eye to accuracy. Were the characters portrayed in the movie as depicted in the book? Did they add characters or change the story? Discuss reasons why the film makers may have chosen to adapt the story in such a way. Finally, discuss which students enjoyed most - the movie or the book - and why.

You will be reminded to compare movies to the book at the end of this unit.

Week 28

Around the World in 80 Days

Chapters 1-4

British Isles (OMB 18)
- Draw boundaries of the United Kingdom and label each country.
- Label London, Liverpool, and Glasgow.

Europe (OMB 25)
- Label England, London, France, Paris.

World (OMB 50)
- Track the "round the world" estimate from the Daily Telegraph in Chapter 4. Mark each place on the world map and draw a green dotted line on the proposed route. On the dotted line, write the time estimate from the newspaper article.
- Follow the path of Phileas Fogg's journey each day by marking it on your world map with a red dotted line.

Trail Blazing

Arts/Crafts
Select from the list in the instructions on page 87.

Chapter 1 - In which Phileas Fogg and Passepartout accept each other: the one as master, the other as man
Describe Phileas Fogg with a focus on his habits and his character. Where did he get the ice that cooled his beverages? Why had he dismissed his servant? What kind of man was the new servant, Passepartout? Do you think he will be better suited to Mr. Fogg's lifestyle? Why or why not?

Additional Research
- Learn more about the 1800s. Compare life in England to life in America at that time.
- Find information about London from a travel video or the Internet.
- What countries make up the United Kingdom, and what is its history?
- Study thermometers; what kind was used in the 1800s? What difference, if any, did they have to our thermometers of today?

Spelling and Vocabulary

enigmatically	taciturn	avaricious
valet	benevolent	entomologist

Chapter 2 - In which Passepartout is convinced that he has at last found his ideal
What did Passepartout do after he was left alone in the house at Savile Row? In what ways were Passepartout and Phileas Fogg alike and in what ways were they different? At what time each evening did Mr. Fogg return home?

Additional Research
- Study Paris, France.
- Research the inner workings of various kinds of clocks; use illustrations.

Spelling and Vocabulary

flurried	physiognomies	rubicund
toilet (as used in the text)	remonstrance	phlegmatic

Chapter 3 - In which a conversation takes place which seems likely to cost Phileas Fogg dear
What foods did Phileas Fogg eat for breakfast? What did he do with his time all day? How did the bank robber get away with his theft so easily? Describe the thief. What did Phileas Fogg wager that he could accomplish, and how much money did he put at risk?

Additional Research

- Learn about different modes of transportation, especially steamers and railroads.
- Start a newspaper unit:
  - Dissect the newspaper with the students, explaining the various sections.
  - Go on a tour of the local newspaper building or have a journalist come to the classroom.
  - Divide children into sections of a newspaper (national, local, lifestyles, etc.) to create their own mini-newspaper about school and local happenings. Make copies to sell or give.
  - OR have students write an article about the bank robbery for the Daily Telegraph.

Math
What is the current exchange rate for British pounds? What does 55,000 pounds equal in today's American dollars? If all the money were to be found, what would be the total of the reward in pounds and in dollars? (Hint: the reward was 2,000 pounds plus five percent of the amount recovered.)

Spelling and Vocabulary

edifice	flunky	shilling sixpence		wager
stoical	composure	expend	antagonists	agitate

Chapter 4 - In which Phileas Fogg astounds Passepartout, his servant
What astounded Passepartout? What time did Phileas Fogg leave the Reform Club, and what time did he depart for his journey? What clothing did they pack? How much money did Fogg take in the carpetbag? What did Phileas Fogg do with the 20 guineas he won at whist? What form of transportation did Phileas Fogg and his companion take to begin their journey around the world? (cab to the train station and train to their first stop in Paris, France) When were they to return to London in order to win the bet?

Additional Research

- Learn about the new Euro and its effect upon the nations that participate.
- How is whist played? In the game of whist, what is a rubber?

Spelling and Vocabulary

guineas	domestic	ensconced	stupefaction

Week 29

Around the World in 80 Days

Chapters 5-8

Mapping and Geography

Mediterranean Sea (OMB 35)

- Label Italy and the cities of Turin and Brindisi in Italy.

Middle East (OMB 37)

- Label Egypt, the Suez Canal, Red Sea, and the Gulf of Aden.
- Label Aden (in Yemen).

World

- Continue to follow the path of Phileas Fogg by marking it on your world map with a red dotted line.

List any geography terms used in the readings this week
List any plants and animals used in the readings this week.

Trail Blazing

Arts/Crafts

Select from the list in the instructions on page 87.

Chapter 5 - In which a new species of funds, unknown to the moneyed men, appears on 'Change

What was the response of the members of the Reform Club and the people in London to the "tour of the world"? What was one of the pet subjects of the English? What incident ended any wager in favor of Phileas Fogg's success?

Spelling and Vocabulary

premium elude dwindle

Chapter 6 - In which Fix, the detective, betrays a very natural impatience

Who was Fix and what was his mission? How long was the Magnolia to remain at Suez to load on coal? How many miles from Suez to Aden?

Additional Research

- The Suez Canal shortened the traveling distance from England to India by about one half. Before its construction, ships had to travel around the Cape of Good Hope. Locate the Cape of Good Hope on your world map. Learn more about canals.
- The steamer, Magnolia, was "built of iron, of two thousand eight hundred tons burden, and five hundred horsepower." Learn more about steamers.

Spelling and Vocabulary

consul contemptible quay

Chapter 7 - Which once more demonstrates the uselessness of passports as aids to detectives
Why did Passepartout bring Phileas Fogg's passport to the consulate? Who kept a notebook recording the time and location of each leg of the journey? Were they on schedule at this point in the trip? Did Phileas Fogg tour the town when he arrived?

Additional Research
• Learn how to obtain a passport. Get a copy of the application from your post office and fill it out as if you were applying for one yourself. Attach a school picture to it if you have extras. Ask someone with a passport to show you theirs. Read the oath imprinted on the passport near the signature.

Spelling and Vocabulary
rogue queried wont

Chapter 8 - In which Passepartout talks rather more, perhaps, than is prudent
To whom did Passepartout talk about the trip? Do you think he should have told a stranger so many details about his master? How long was the trip from Suez to Bombay, India, supposed to take? Why didn't Fix arrest Phileas Fogg?

Additional Research
• Learn about telegraphs.
• Learn about time zones.

Spelling and Vocabulary
volubly chronometer fob cogitating

Week 30

Around the World in 80 Days

Chapters 9-12

Mapping and Geography

Middle East

- Label the Strait of Bab-el-Mandeb.

India (OMB 29)

- Label the following cities: Calcutta, Bombay, Madras, Agra, Allahabad, and Banaras (Varanasi).
- Label the Indus and Ganges Rivers.
- Draw the path of the Great Indian Peninsula Railway as given in the story. (Some of the places are hard to find, but see what you can do with this challenging project!)

through Bombay --> to Callyan (Kalyan) --> to Pounah (Poona) -->
past Nassik (Nasik) --> to Berhampoor (perhaps Burhanpur -on the Tapi River east of Surat) -->
to Allahabad --> eastward meeting the Ganges River at Benares -->
southeastward by Burdivan (Burdwan) ending in Calcutta

World

- Continue to follow the path of Phileas Fogg by marking it on your world map with a red dotted line.

List any geography terms used in the readings this week.
List any plants and animals used in the readings this week.

Trail Blazing

Arts/Crafts

Select from the list in the instructions on page 87.

Chapter 9 - In which the Red Sea and the Indian Ocean prove propitious to the designs of Phileas Fogg

How far was the trip from Suez to Aden, and how much time was allocated to make the trip? (1310 miles; 138 hours) What does Phileas Fogg do to pass the time? There were 25,000 inhabitants of many cultures in Aden. Name six of them. How many hours expected to reach Bombay? When the steamer arrives in Bombay, was it on schedule?

Math

It costs the Peninsular Company 800,000 pounds a year to fuel its steamers. How much is this in today's American dollars?

Additional Research

- Study Egyptian history and culture.
- Begin a study of India. Include the animals in your study. Plan to write a report or make a notebook of your findings on this nation.

Spelling and Vocabulary

itinerary boisterous delusion pagodas

Chapter 10 - In which Passepartout is only too glad to get off with the loss of his shoes
What forbidden act did Passepartout unknowingly commit? What common house pet was once considered sacred in India? What happened to Passepartout's shoes? On what form of transportation did Fogg depart the city of Bombay?

Additional Research

- Add to your study of India information about its religious beliefs. Learn about the agriculture of India, including dates, coffee, and spices.

Spelling and Vocabulary

dominion rajahs palanquin rogue promenade

Chapter 11 - In which Phileas Fogg secures a curious means of conveyance at a fabulous price
Passepartout began to take a different perspective on this journey; can you describe the change? There were no tracks between Kholby and Allahabad. How did Phileas Fogg plan to continue to Allahabad and what did it cost him? Who will take this leg of the journey with him? What time did the train stop, and what time did they resume their journey? Were they on schedule?

Additional Research
- Learn about plantations.
- Study Indian architecture and add to your report on India.
- Opium and indigo merchants were on the train. Learn more about these two products. How were they used?

Spelling and Vocabulary
opium indigo terrestrial eccentric
viaducts minarets acacias conveyance

Chapter 12 - In which Phileas Fogg and his companions venture across the Indian forests, and what ensued
What was the name of the elephant? How smooth was the traveling experience? What was the suttee? If the sacrifice was not a voluntary one, why didn't the woman struggle against it? Whose idea was it to save the woman? What does this tell you about Phileas Fogg's character?

Additional Research
- Add to your study of India information about its musical instruments.
- Study the training of elephants.
- Learn about the beliefs of the Hindu religion.
- Learn about the igneous rock syenite.

Spelling and Vocabulary
howdah phlegm Parsee lugubrious
caparisoned zebu damascened

Week 31

Around the World in 80 Days

Chapters 13-16

Mapping and Geography

Asia

- Find and label Calcutta, Hong Kong, and the Andaman Islands.
- Label the Bay of Bengal and the Indian Ocean.
- Label China and Japan.

World

- Continue to follow the path of Phileas Fogg each week by marking it on your world map with a red dotted line.

List any geography terms used in the readings this week.
List any plants and animals used in the readings this week.

Trail Blazing

Arts/Crafts

Select from the list in the instructions on page 87.

Chapter 13 - In which Passepartout receives a new proof that fortune favours the brave

If they were caught, what could be the punishment for attempting this rescue? What kind of background did Aouda have? Why did Sir Frances and the Parsee guide give up on the plan to rescue Aouda during the night? Who actually rescued her, and how did he do it? What does this show you about his character?

Additional Research

- Study Indian religious beliefs.
- Learn about the phases of the moon.

Spelling and Vocabulary

rajah sabers anon wane

Chapter 14 - In which Phileas Fogg descends the whole length of the beautiful valley of the Ganges without ever thinking of seeing it

How do you think Passepartout's background helped him pull off this daring rescue? When did they reach the train station at Allahabad? How did Phileas Fogg compensate the guide for his devotion? Why did they decide to take Aouda out of India with them, and where would she go from there? What nickname did the Orientals have for Benares? Who parted company with the group at Benares? When arriving in Calcutta after all of this adventure, was Phileas Fogg behind schedule?

Additional Research

- Add to your study of India. Don't forget to include the animals from this chapter.
- Learn about iron foundries.
- Study the growth and processing of corn, wheat, and barley.

Spelling and Vocabulary

stupefied foundries verdure venerated

Chapter 15 - In which the bag of bank notes disgorges some thousands of pounds more
What did Passepartout think they had done wrong to be arrested? Why were they really apprehended? How did they gain the freedom to proceed on their journey? Why had Fix remained behind when Phileas Fogg boarded the train in Bombay? Did these events seem to disturb Mr. Fogg? What is the name of the ship they will take from Calcutta to Hong Kong?

Additional Research
• Continue your study of India and begin to finalize your research report or organize your India notebook.

Spelling and Vocabulary
disgorge crestfallen conjecture

Chapter 16 - In which Fix does not seem to understand in the least what is said to him
How long was the trip from Calcutta to Hong Kong? With the many conversations and time spent with Aouda, did Phileas Fogg become smitten with her or change his demeanor in any way? What went on between Passepartout and Fix? What did Fix hope to be able to do when they arrive in Hong Kong?

Additional Research
• Learn about Morse Code. Put it on a 3x5 index card and begin to learn it for yourself. Practice tapping short messages to a friend or sibling.

Spelling and Vocabulary
Create a word search puzzle from previous words and have others complete it.

Week 32

Around the World in 80 Days

Chapters 17-20

Mapping and Geography

Asia

- Label the Strait of Malacca, Sumatra, and Singapore.
- Label the Pacific Ocean.
- Label Yokohama, Nagasaki, and Shanghai.

World

- Continue to follow the path of Phileas Fogg by marking it on your world map with a red dotted line.

List any geography terms used in the readings this week
List any plants and animals used in the readings this week.

Trail Blazing

Arts/Crafts

Select from the list in the instructions on page 87.

Chapter 17 - Showing what happened on the voyage from Singapore to Hong Kong
Passepartout began to suspect Fix of hiding something. What story did he make up in his mind for the peculiar appearance of Fix on so many parts of their voyage? Contrast the attitudes and reaction of Phileas Fogg and Passepartout to the possibility of getting behind schedule.

Additional Research

- Study the Chinese culture and life in China. Include the plants and animals from this chapter, as well as clothing, foods, and Chinese traditions.

Spelling and Vocabulary

cudgeled

Chapter 18 - In which Phileas Fogg, Passepartout and Fix go each about his business
What threatened to delay the journey? Who was most impatient with the delays and the weather? What does it mean that Mr. Fogg moves mathematically? What did Phileas Fogg discover about Aouda's relatives at the Exchange? What assignment did Phileas Fogg give to his servant Passepartout?

Additional Research

- Learn about compasses
- Study reefs and their effect on sea life and on ships.

Spelling and Vocabulary

tempestuous propitious flotilla squall

Chapter 19 - In which Passepartout takes a too great interest in his master, and what comes of it

Why were some older men dressed in yellow? Why did Fix wish to detain Fogg in Hong Kong? How did Fix prevent Phileas Fogg from boarding the Carnatic on time? Passepartout was staunchly loyal to his master, but made some bad choices that day. What were some consequences of his choices? Discuss the character of a man who would leave someone he had befriended in a saloon in a foreign country, in the condition Fix left Passepartout.

Additional Research

• Continue your study of China.

Spelling and Vocabulary

antipodes parterres inveigh raillery

Chapter 20 - In which Fix comes face to face with Phileas Fogg

Why do you suppose Mr. Fogg often responded to delays and circumstances by the phrase, "It is in the interest of my journey - a part of my programme"? How did Phileas Fogg react to the news that he missed the sailing of the Carnatic? Who was most disturbed at the disappearance of Passepartout? How did Mr.Fogg solve his problem of missing the boat? Who joined Mr. Fogg and Aouda on the Tankadere? What was Fix's reaction to Passepartout's apparent disappearance? Is he to be trusted?

Additional Research

• Continue with any previous assignments.

Spelling and Vocabulary

remonstrance divan

Around the World in 80 Days

Chapters 21-24

Mapping and Geography

Asia
• Mark and label the Tropic of Cancer.

World
• Continue to follow the path of Phileas Fogg by marking it on your world map with a red dotted line.

List any geography terms used in the readings this week
List any plants and animals used in the readings this week.

Trail Blazing

Arts/Crafts
Select from the list in the instructions on page 87.

Chapter 21 - In which the master of the Tankadere runs great risk of losing a reward of two hundred pounds
How many miles was this voyage to Shanghai? Where were Fogg and his companions sailing? What will happen to Fix's warrant when Mr. Fogg arrives in Japan? What time did Mr. Fogg descend to his room? (midnight just like when he was home) Did Fix ever feel guilty for traveling at the expense of the man he intended to arrest? How did Mr. Fogg respond to the typhoon? Why did they fire a cannon?

Additional Research
• Study weather and climate. What defines a typhoon?
• Learn about the moon's phases and equinoxes.
• Learn more about sailing vessels.

Spelling and Vocabulary

equinoxes	bow	regatta	adroit
gale	barometer	tempest	typhoon

Chapter 22 - In which Passepartout find out that, even at the antipodes, it is convenient to have some money in one's pocket

Passepartout made it onto the Carnatic on time, but how far away from China had it sailed before he woke up? Did he realize his master was not on board? Was it common to smoke opium in Japan as it was in China? What bird did the Japanese consider to be sacred? Why?

Additional Research
• Learn about Japan. Include its plants, birds, and foods.
• Study the effects of narcotics on the body.

Spelling and Vocabulary

torpor	inveigled	peristyle	importunate

Chapter 23 - In which Passepartout's nose becomes outrageously long

How did Passepartout first determine to get some money? What kind of job did Passepartout get? How did his past prepare him to perform such a routine? How long was the false nose Passepartout had to wear? How did the pyramid come crashing down, who caused it to happen, and why? At what time did Mr. Fogg, Aouda, and Passepartout board the ship for America?

Additional Research
- Learn more about different circuses.
- Research the history of Japanese clothing.

Spelling and Vocabulary
accoutered grimaces equilibrist

Chapter 24 - During which Mr. Fogg and party cross the Pacific Ocean

When Mr. Fogg discovered a Frenchman had arrived on the Carnatic, were you surprised that he went in search of his servant? What was the name of the steamer they would take to America and what kind of ship was it? What delighted Passepartout on November 23? Why did Fix decide to help Fogg and no longer hinder his efforts? When did the General Grant arrive in San Francisco?

Math
How much extra did Mr. Fogg pay John Bunsby for getting him to Shanghai? How much is that in American dollars?

Additional Research
- Learn about meridians and parallels.
- Learn about paddleboats.

Spelling and Vocabulary
pugilistic treason

Week 34

Around the World in 80 Days

Chapters 25-28

Mapping and Geography

United States (OMB 60)

- Label San Francisco, Sacramento, Omaha, New York, Salt Lake City, Kansas, Colorado, Oregon, Missouri, and Utah.
- Shade the Rocky Mountains and the Sierra Nevada Mountains lightly in brown and label.
- Label the Great Salt Lakes, Platte River, and Arkansas River.
- Draw the route taken by the Central Pacific Railroad. This will be challenging, as not all places described are city names. Use a historical atlas of the U.S. for a more accurate map of the railroad:

Oakland, California --> Sacramento, California --> Auburn, California through Carson Valley --> Reno, Nevada --> along theHumboldt River --> the Great Salt Lake, Utah northwest border --> Ogden, Utah --> northward to Weber River --> east toward Wasatch Mountains --> follow the valley of Bitter Creek to Fort Bridger --> Wyoming territory at the Green River station --> North Platte River --> Medicine Bow

World

- Continue to follow the path of Phileas Fogg by marking it on your world map with a red dotted line.

List any geography terms used in the readings this week.
List any plants and animals used in the readings this week.

Trail Blazing

Arts/Crafts

Select from the list in the instructions on page 87.

Chapter 25 - In which a slight glimpse is had of San Francisco
What did the International Hotel resemble to Passepartout? What food was given away free at the hotel? Did Mr. Fogg know about the way Fix treated Passepartout? How did Fix get a black eye?

Additional Research

- Study political campaigns; how have they changed in the past 100 years? What methods do politicians use to compel people to vote?
- Study the Native Americans, especially the Sioux and Pawnee.
- Learn more about the architecture of an Anglo-Saxon Gothic church. Obtain pictures to see this style.

Spelling and Vocabulary
Review words already studied.

Chapter 26 - In which Phileas Fogg and party travel by the Pacific Ocean
How many miles of railroad connects New York and San Francisco? (3786) How long did it take to travel that distance before the railroad was established and how long after? (six months/seven days) When did the train leave Oakland Station? How was the train transformed in the evening? What unexpectedly stopped the train in Nevada for three hours?

Math
Calculate the average miles per hour traveled from the information given in this chapter.

Additional Research
• Learn about the intercontinental railroad.

Spelling and Vocabulary
ruminating

Chapter 27 - In which Passepartout undergoes, at a speed of twenty miles an hour, a course of Mormon history
How many came to hear Elder Hitch's message on Mormonism? How many remained until the end of his message? After listening patiently to Elder Hitch did Passepartout agree to join him? When did the train reach Ogden? In what American town did they spend two hours?

Additional Research
• Learn about the Great Salt Lake in Utah.
• Study the layout of American towns.

Spelling and Vocabulary
cravat polygamy

Chapter 28 - In which Passepartout does not succeed in making anybody listen to reason
Who unexpectedly appeared on the train? What did Mr. Fogg's companions do to prevent him from knowing? What was the solution to crossing the bridge? Did Passepartout have a good idea? Why do you think no one listened to him?

Additional Research
• Study suspension bridges.

Spelling and Vocabulary
Review spelling of all locations (countries, cities, rivers, mountains) previously studied.

Week 35

Around the World in 80 Days

Chapters 29-32

Mapping and Geography

United States

- Find and label these states: Wyoming, Colorado, Nebraska, Iowa, Illinois, Indiana, Ohio, Pennsylvania, and New Jersey.
- Draw and label these bodies of water: Hudson River, Lake Michigan.
- Find and label the 101st meridian, Jersey City, and Long Island.
- Continue to draw the route taken by the Central Pacific Railroad:

Fort Saunders	-->	crossing Cheyenne Pass	-->	to Evans Pass	-->
Denver	-->	Camp Walbach	-->	Lodge Pole Creek	-->
Nebraska near Sedgwick	-->	Julesburg on southern branch of the Platte River			-->
Fort McPherson	-->	North Platte, Nebraska	-->	101st meridian	-->
Ft. Kearney	-->	Omaha, Nebraska	-->	Council Bluffs, Iowa	-->
Des Moines, Iowa	-->	Iowa City, Iowa	-->	across the Mississippi at	
Davenport, IA	-->	Rock Island to Illinois	-->	Chicago, Ilinois	

- Draw the route of the Pittsburgh, Fort Wayne, and Chicago Railway across Indiana, Ohio, Pennsylvania, and New Jersey. Use a different color for this railway.

World

Continue to follow the path of Phileas Fogg by marking it on your world map with a red dotted line. The story doesn't give much detail. You may need to consult an encyclopedia or historical atlas for the train route.

Trail Blazing

Arts/Crafts

Select from the list in the instructions on page 87.

Chapter 29 - In which certain incidents are narrated which are only to be met with on American railroads

How did Mr. Fogg and Colonel Proctor decide to settle their difference with "honor"? Why were they going to perform the duel on the train? What happened to stop the duel before it started? How did Passepartout's gymnastic background aid him in stopping the train? What Indians attacked the train? Who was among the missing at the close of this chapter?

Additional Research
- Learn more about the Sioux and Pawnee Indians.

Spelling and Vocabulary

alacrity clamorous waylaid

Chapter 30 - In which Phileas Fogg simply does his duty

What was it that Phileas Fogg decided was his duty? Did he show frustration with the delay in his journey? What fort provided safety for the passengers? Why do you suppose the captain changed his mind about sending out a search party for the missing people? How much money did Fogg offer to split among those who helped? Did they agree to help before he offered the money, or after? (Did you notice he offered dollars, not pounds? Were they interchangeable back then, or do you suppose it is an oversight of the author?) Why did Mr. Fix decline to board the train? Was this in keeping with his character?

Additional Research

- Continue research on Native Americans. What were the distinguishing characteristics between the different nations? Why would the Sioux attack this train?
- Learn about Fort Kearney and its role in American history.

Spelling and Vocabulary

succor reconnaissance

Chapter 31 - In which Fix the detective considerably furthers the interests of Phileas Fogg

At twenty hours behind schedule who offered a solution? Who was Mudge? How far is the trip from Fort Kearny to Omaha? (200 miles) What powered the sledge? Did they arrive on time to catch the train to New York? Did they arrive in New York in time to catch the China bound for Liverpool?

Additional Research

- Study railroad history (especially in America).
- Learn more about wind power.
- Learn more about a sledge and how it is used. Obtain pictures if possible.

Spelling and Vocabulary

Review previous words.

Chapter 32 - In which Phileas Fogg engages in a direct struggle with bad fortune

What was Mr. Fogg's initial response to the news that he had missed the China and no other ship was available to get him to Liverpool on time? Was this a predictable response? Mr. Fogg arranged to sail on the Henrietta; where was it bound? How much time did he have between his agreement with the captain and the time of departure? Why was Fix concerned at the continued expenses being drained from the alleged stolen money?

Spelling and Vocabulary

espied disembark gamut indemnified

Week 36

Around the World in 80 Days

Chapters 33-37

Mapping and Geography

British Isles

• Find and label Newfoundland, Bordeaux, Queenstown, and Edinburgh (Scotland).

World

• Continue to follow the path of Phileas Fogg by marking it on your world map with a red dotted line.

Chapter 33 - In which Phileas Fogg shows himself equal to the occasion

What was the captain's name, and where did he spend the majority of the trip? What was Mr. Fogg's deception? What skill do we learn Mr. Fogg has? Why was there not sufficient coal on board to complete the trip to Liverpool? How did Mr. Fogg solve his problem? Who became interested in Mr. Fogg's project? What did Fix do when Phileas Fogg disembarked on the Liverpool quay?

Additional Research

• Learn ship terms (knots, keel, masts, poop, rafts, spars, fittings, and topsides).

Spelling and Vocabulary

loquacious	dexterous	conjecture
presentiment	epithet	colloquy

Chapter 34 - In which Phileas Fogg at last reaches London

Why was Phileas Fogg arrested and why was he released? Did he ever show emotion at the apparent loss of his wager? Who seemed most disappointed? Why did Passepartout feel responsible? Phileas Fogg finally showed emotion when he was released; what did he do? When did they arrive in London?

Additional Research

• Learn about English culture. Study their police, laws, foods, and custom houses.
• Play the board game Scotland Yard, if available.

Spelling and Vocabulary

Review previous lessons.

Chapter 35 - In which Phileas Fogg does not have to repeat his orders to Passepartout twice

Mr. Fogg never seemed to be angry or frustrated at the troubles caused by Passepartout; how would you have handled these situations? Having lost nearly all he owned, for whom did Phileas Fogg take responsibility? In the entire story Mr. Fogg never outwardly displays love for Aouda; were you surprised when he accepted her proposal of marriage with, "Yes, by all that is holiest, I love you, and I am entirely yours!"? Why or why not?

Additional Research

• Complete any research or additional studies.

Spelling and Vocabulary

domicile	imperturbable	tranquility
expedient	pensive	

Chapter 36 - In which Phileas Fogg's name is once more at a premium on 'Change
Why did the members of the Reform Club renew their interest in Phileas Fogg's journey and begin wagering again on its success or failure? Who had the most confidence in Phileas Fogg to accomplish this task so associated with keeping a strict time schedule? How much time did Mr. Fogg have to spare when he entered the Reform Club?

Spelling and Vocabulary
antagonist eccentric

Chapter 37 - In which it is shown that Phileas Fogg gained nothing by his tour around the world, unless it were happiness
Did you suspect they were a day ahead of schedule before it was revealed in the story?

Additional Research
- Study the history and science of calculating earth's degrees and the development of the calendar including leap year.

Spelling and Vocabulary
fastidious circumference pecuniary conveyance

WRAPPING IT UP

You've just completed this Geography through Literature unit. Finalize your study by discussing the book. What conclusions can you draw from this story about:
- managing your time?
- gambling?

Which character are you most like? Who do you most admire? Why?
For each of the main characters - Fogg, Passepartout, Fix, and Aouda:
- Give three character traits.
- Describe each in one word.

Before reading the book did you think the trip around the world occurred in a hot air balloon? Watch various video movie presentations of *Around the World in 80 Days*. How closely does each movie follow the book? How well did they portray the characters? Discuss why movie producers would change the story.

Are you interested in reading other books by Jules Verne?

Appendix

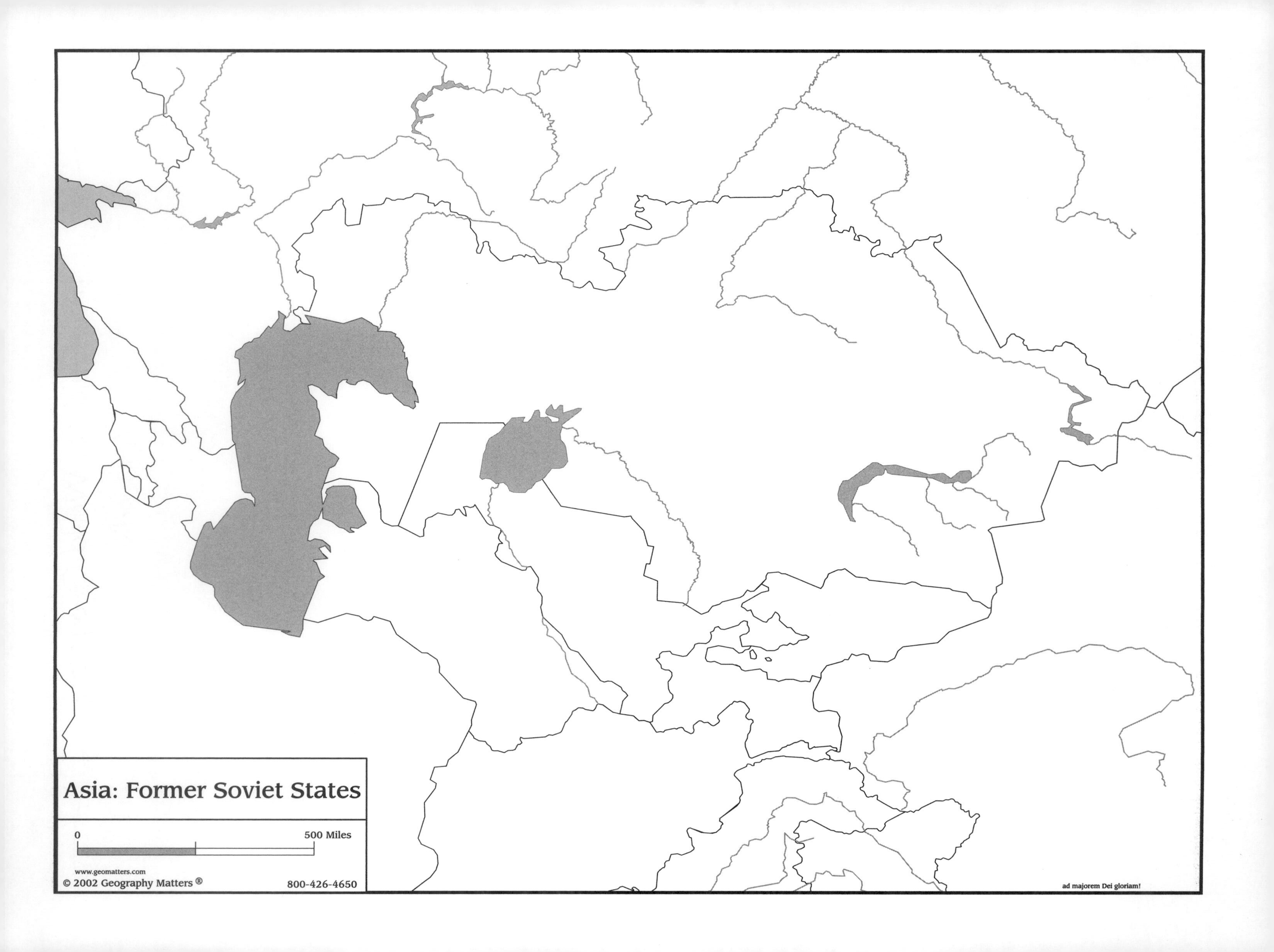
Asia: Former Soviet States
0
500 Miles
www.geomatters.com
© 2002 Geography Matters ®
800-426-4650
ad majorem Dei gloriam!

England:
The Whipping Boy
The Big Concrete Lorry
Secret Garden

Netherlands:
The Wheel on the School
The Hole in the Dike

Sweden:
Pippi Longstocking
Pelle's New Suit

France:
Twenty and Ten
Madeline

Italy:
Haydn, Mozart, Schubert
Pinocchio

Russia:
Endless Steppe
Peter and the Wolf

Scotland:
John Paul Jones
Always Room for One More

Spain:
Ferdinand

Africa:
The Voyages of Doctor Doolittle
Bringing the Rain to Kapiti Plain

Israel:
Frank Peretti Series
Joseph and His Magnificent Coat

Asia:
House of Sixty Fathers
Ping

Pacific Ocean:
Call It Courage

Indonesia:
The Twenty-One Balloons
The Little Princess

Canada:
Three Days in a Canoe
Petite Suzanne
Hatchet

South America:
The Amazon Stranger
Moon Rope

Atlantic Ocean:
Mandie and the Ship Board Mystery
Tim in Danger
Titanic

Geography Through Literature Book List for Older Students

North America:

My Indian Boyhood
Paddle-to-the-Sea
The Tree in the Trail
Stories of Great Americans
Journey to Jericho
Thunder Rolling in the Mountains
World of the American Indians
Undaunted Courage
James Michener novels

South America:

Secret of the Andes
Treasure of the Incas: A Tale of Adventure in Peru
The Fate of the Yellow Woodbee
Simon Bolivar the Great Liberator
A Long Vacation
Night Flight
It's a Jungle out There!
Green Mansions
Far Away and Long Ago
Through Gates of Splendor

Europe:

Prince Henry the Navigator
In Freedom's Cause: A Story of Wallace and Bruce
Puppy Lost in Lapland
Snow Treasure
Kidnapped
Under Wellington's Command: A Tale of the Peninsular War
Chronicle of Conquest of Granada
Don Quixote
Henry the Navigator

Africa:

King of the Wind
The Barbary Pirates
Star of Light
King Solomon's Mines
Born Free
The Covenant
My African Journey

Asia:

Gengis Khan - Mongol Emperor
Through Russian Snows
Silk Route
Vinegar Boy
Adventures and Discoveries of Marco Polo
Lawrence of Arabia
Endless Steppe
The Talisman
The Bronze Bow
Dancing Bear
Anna and the King
Zion Chronicles
Michael Strogoff: A Courier of the Czar
Doctor Zhivago
Anna Karenina
Ben Hur
Red Storm Rising

Oceania and Pacific Islands:

Red Earth, Blue Sky
Maori and Settler: A Story of the New Zealand War
Boy Alone
The Remarkable Voyages of Captain Cook
Island of the Blue Dolphins
Swiss Family Robinson
The New Atlantis
Hawaii (by James Michener)

Illustrated Geography Dictionary

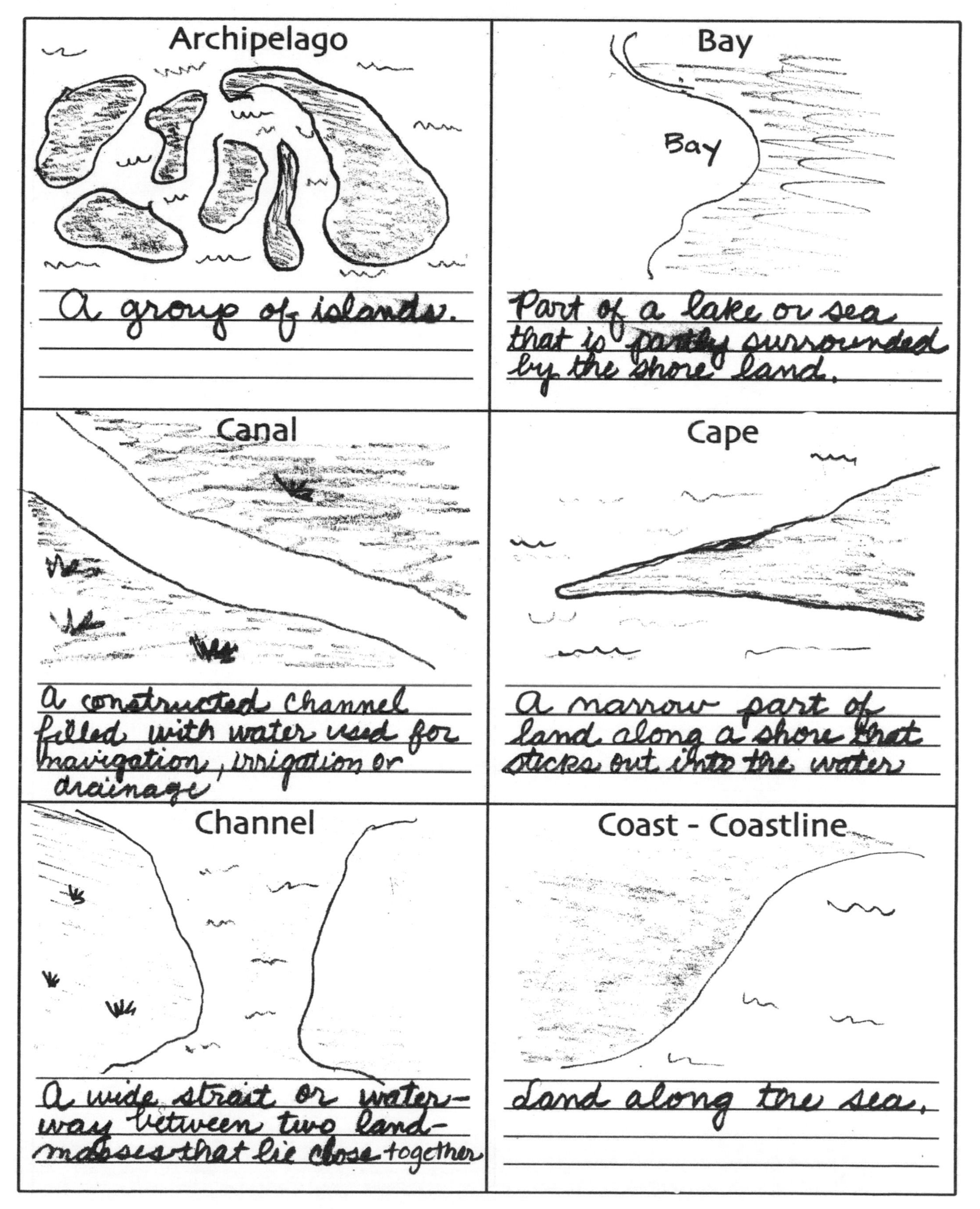

archipelago	bay
canal	cape
channel	coast - coastline

delta	desert
gulf	harbor
hemisphere	island

latitude - parallels	peninsula
plain	plateau
river	savanna

sea	sound
strait	steppe
trench	volcano

Illustrated Geography Dictionary

ANSWERS

Week 1

1-model of the earth; sphere (or geoid)
2-continents; ocean
3-North America, South America, Antarctica, Europe, Asia, Africa, Oceania (including Australia); answer varies
4-lines of latitude; equator

1-precipitation and temperature; atlas
2-Asia, Africa, North America, South America, Antarctica, Europe, Oceania (including Australia); compass rose
3-lines of latitude; North America, Europe, Asia, Africa
4-North Pole; North Pole

1-7926 miles; the moon
2-area given in million square miles: Asia 17.3, Africa 11.7, North America 9.5, South America 6.9, Antarctica 5.4, Europe 3.8, Oceania, including Australia 3.3; 57.9 million square miles
3-lines of latitude; South America, Antarctica, Australia, Africa
4-Arctic Circle; 66.5°

*Week 2

1-Atlantic Ocean, Pacific Ocean, Indian Ocean, Southern Ocean, Arctic Ocean; Atlantic Ocean
2-lines of longitude; prime meridian
3-curvy blue lines; blue
4-Pacific Ocean, Indian Ocean; South America

1-Pacific Ocean, Atlantic Ocean, Indian Ocean Southern Ocean, Arctic Ocean; Indian, Atlantic, Pacific, and Southern Occans
2-Europe, Asia, Africa, Australia; North America South America
3-3%; political maps
4-Atlantic, Pacific, Arctic; physical maps

1-*area in million sqare miles: Pacific 63.8, Atlantic 31.8, Indian 28.9, Arctic 5.4; total is about 130 million sqare miles
2-Greenwich, England; 75%, 97%
3-plate tectonics; volcanoes and earthquakes
4-solid inner core 1520 miles, molten outer core 1405 miles lower mantle 1365 miles, upper mantle 415 miles, solid crust 0-9 miles; 24,900 miles

Week 3

1-Northern and Western; Canada
2-island; Arctic Ocean, Pacific Ocean, Atlantic Ocean
3-Canada, Greenland, and U.S.; Canada
4-Bering Sea; Asia

1-Canada, U.S., Mexico; U.S.
2-Greenland; 840,000 sq. mi;
3-Nunavut; a three-sided piece of land jutting out into a large body of water
4-66.5°; a sea

1-Pacific Ocean; distance scale
2-channel; Great Bear Lake
3-Gerardus Mercator in 1589; Mexico City
4-New York City; 79%

***Week 2 secondary note re: oceans**
As of this printing the *Answer Atlas* did not list the Southern Ocean on its chart of information, therefore it is not included in the answer key.

ANSWERS

Week 4

1-Canada; Great Plains
2-Rio Grande; wheat
3-Pacific Ocean; Canada and U.S. (Alaska)
4-Hudson Bay; Canada

1-The larger islands include any four of the following: Banks, Victoria, Baffin, Devon, Melville, Prince of Wales, and Somerset Islands; Beaufort Sea
2-Canada and U.S.; Greenland and Baffin Island
3-Mississippi River; hunting, forestry,and subsistence farming (see thematic map)
4-Aleutian Islands; Appalachian Mountains

1-French, British, and Native Americans; Manitoba, Ontario, Quebec, and Nunavut
2-east; Yukon, Northwest, and Nunavut
3-Viscount Melville Sound; British Columbia, Alberta, Saskatchewan, Manitoba, Ontario, Quebec, Newfoundland and Labrador, New Brunswick, Nova Scotia, and Prince Edward Island
4-Alaska, Hawaii, Washington, and California; on the Alaskan Peninsula and the Aleutian Islands

Week 5

1-no; Appalachian Mountains
2-Mt. McKinley; Alaska, U.S.
3-five; a triangle
4-Pierre; west

1-Washington; eastern and central United States
2-Lake Superior; Lake Michigan
3-Mississippi-Missouri river system (or Mississippi River); legend
4-Louisiana; Mount McKinley,20,320 ft.

1-U.S.; Mexico City
2-Mississippi-Missouri; Death Valley, California (elevation 282 ft. below sea level)
3-Continental Divide; North American Free Trade Agreement, enacted to remove all trade restrictions between Canada, U.S., and Mexico
4-trench; mountains and widely spaced mountains (or whatever term your atlas uses for that area)

Week 6

1-gulf; Gulf of Mexico
2-Mississippi River; Mexico City
3-Rio Grande; map
4-Mexico; north

1-Cuba, Jamaica, Haiti, Dominican Republic; Cuba
2-Baja California, Yucatan Peninsula, the U.S. state of Florida; Nicaragua
3-Yucatan Channel; Texas, Louisiana, Mississippi, Alabama, Florida
4-San Antonio; San Diego

1-Yucatan Channel; Greater Antilles and Lesser Antilles
2-above; west
3-Puerto Rico Trench; Guadalajara
4-plateau; Tropic of Capricorn at 23.5°S and Tropic of Cancer at 23.5°N

ANSWERS

Week 7

1-Guatemala; Nicaragua, Costa Rica, and Panama
2-Mexico and El Salvador; Panama Canal
3-Carribbean Sea; Costa Rica
4-north; Havana

1-Belize, Guatemala, Honduras, El Salvador, Nicaragua, Costa Rica, and Panama; West Indies
2-the Straits of Florida; El Salvador and Nicaragua
3-Mexico City; Atlantic Ocean and Pacific Ocean
4-Jamaica; hot and rainy

1-southern boundary of Mexico and northern boundary of Colombia; Belize, Guatemala, Honduras, El Salvador, Nicaragua, Costa Rica, and Panama
2-Isthmus of Panama (Istmo de Panama); sound
3-Trinidad, it exports about 80% of its oil; inlet
4-Lake Nicaragua; Middle America

Week 8

1-Brazil; capital city
2-Brazil and Colombia; Colombia
3-western; Chile and Argentina
4-Bolivia and Paraguay; Peru

1-Answers in this order: French Guiana, Suriname, Guyana, Venezuela, Columbia, Ecuador, Peru, Bolivia, Chile, Argentina, Uruguay, Paraguay, Brazil; Brazil
2-north and south; archipelago
3-minimum of 100 inches per year; Falkland Islands, United Kingdom
4-Incas; Galapagos Islands

1-Angel Falls (3212 ft); tropical rainforests and tropical savannas
2-Orinoco River; Andes
3-southern part of Argentina; Argentina - highest: Cerro Aconcagua; lowest: Laguna del Carbón; hottest: Rivadavia; coldest: Sarmiento
4-Rio de Janeiro; Atacam Desert

Week 9

1-Brazil; equator
2-Santiago; north
3-rain forest; below
4-Brazil; Caracas

1-Amazon River, Orinoco River, Rio de la Plata River; Amazon
2-Ecuador, Colombia, and Brazil; earthquakes, volcanoes, and tsunamis
3-a stream or river that flows into a larger stream or river; Brazil
4-Chile, Argentina, Paraguay, and Brazil; Pacific Coast

1-Ushuaia, Argentina; 55°S
2-Cape Horn; 6,180,000
3-Venezuela, Colombia, cuador, Peru, Bolivia, Chile, Argentina; hot and rainy
4-Ecuador; Bolivia and Paraguay

ANSWERS

Week 10

1-Norway and Sweden; Denmark
2-zero; United Kingdom
3-Danube River; Netherlands
4-English Channel; Iceland

1-Norway, United Kingdom, Belgium, Netherlands, Germany, and Denmark; the main rivers are Rhine, Elbe, and Danube
2-North Sea and Atlantic Ocean; Norway, Sweden, Finland, and Russia
3-Scotland, England, Wales, and Northern Ireland; St. George's Channel
4-moderate, mild, rainy; agriculture

1-Scandinavia; Netherlands
2-Lapland; The Hague (in the Netherlands)
3-southern; Baltic Sea
4-United Kingdom and Ireland; Great Britain and Northern Ireland

Note: Great Britain is England, Scotland, and Wales.

Week 11

1-Ural Mountains; Moscow
2-Estonia, Latvia, Lithuania; Belarus
3-Black Sea; mountainous
4-Danube River; Volga River

1-Germany, Poland, Lithuania, Belarus, and Russia; Ural Mountains to the east, Caucasus Mountains to the south
2-Romania; Russia, Ukraine,Romania, Bulgaria, and a part of Turkey
3-Moldova; north
4-Sofia; Vilnius (Lithuania)

1-Kiev (Ukraine); Petrograd and Leningrad
2-Romania; Bulgaria
3-Ural Mountains, Ural River, Caspian Sea, Black Sea, and Caucasus Mountains; Slovakia, Ukraine, and Romania
4-Helsinki (Finland); Ukraine and Romania

Week 12

1-Poland; Prague
2-Danube River; any two of the following: Croatia, Bosnia and Herzegovina, Serbia and Montenegro, and Albania
3-north; any of the following: Vienna, Austria; Budapest, Hungary; Belgrade, Serbia and Montenegro
4-Eastern Hemisphere; compass rose

1-Croatia; Bosnia and Herzegovina
2-Czech Republic; Hungary
3-Poland; Budapest
4-Albania; Macedonia

1-Warsaw, Poland; north
2-Slovakia; Prague, Czech Rep. and Bratislava, Slovakia
3-Croatia, Serbia and Montenegro, Albania; Bosnia and Herzegovina
4-Albania;any three of the following: Bucharest, Romania; Sofia, Bulgaria; Ljubljana, Slovenia; Zagreb, Croatia; or Sarajevo, Bosnia and Herzegovina

ANSWERS

Week 13

1-Spain and Portugal; Paris, France
2-Strait of Gibraltar; Athens, Greece
3-Italy; France, Switzerland, Austria, Slovenia
4-Alps; Matterhorn

1-bay; Spain and France
2-San Marino; Pyrenees Mountains
3-Andorra; Italy
4-Apennines; Vesuvius

1-southeastern France; Athens, Greece
2-Vatican City; swimming
3-France, Switzerland, Austria; Portugal
4-Malta; San Marino

Week 14

1-Mediterranean Sea; Atlas Mountains
2-Somalia; Egypt, Sudan, Eritrea
3-desert; Egypt, Sudan, Uganda
4-Sahara Desert; Chad

1-Morocco, Algeria, Tunisia, Libya, and Egypt; Gulf of Sidra
2-Nile in Egypt; Ras Dejen Mountain elevation above 15,000, Ethiopian Plateau, Rift alley
3-Sahara Desert; Mauritania, Mali, Algeria, Libya, Egypt (not Western Sahara, which is actually a territory of Morocco)
4-11,204 feet; Eritrea, Red Sea

1-Khartoum, Sudan; Cape Ben Sekka in Tunisia
2-Ethiopia and Somalia; oil
3-from south to north; Tripoli, Libya
4-Atlas Mountains; Mali, Niger, Chad

Week 15

1-8 (Sierra Leone, Liberia, Côte d'Ivoire, Ghana, Togo, Benin, Nigeria, South Africa); any three of the following: Western Sahara, Mauritania, Senegal, The Gambia, and Guinea-Bissau
2-Burkina Faso; Nigeria
3-Cape Verde; western
4-herding, hunting, small farming; west

1-Niger, Mali, Burkina Faso; Liberia
2-The Gambia; Cape Verde
3-Sahel; Guinea-Bissau
4-Niger River; Togo

1-Senegal, The Gambia, Guinea- Bissau; Liberia (Monrovia is its capital.)
2-Sahel; Ghana, Togo, Benin,Nigeria, Equatorial Guinea,and Cameroon
3-southern Nigeria at the Atlantic Ocean; Guinea
4-Ouagadougou, Burkina Faso; Ivory Coast

ANSWERS

Week 16

1-Democratic Republic of the Congo (formerly called Zaire); Tanzania
2-east; Congo, Democratic Republic of the Congo, Uganda
3-Cameroon, Central African Republic; Indian Ocean
4-Congo River; lowland gorilla, chimpanzee, mandrill, okapi, among many others

1-savanna(h); Democratic republic of the Congo
2-tropical, hot, with rain; Lake Victoria
3-Great Rift Valley; Tanzania, Kenya, Somalia
4-Tanzania, 19,340 feet
Guinea, Sierra Leone, Liberia, Côte d'Ivoire, Nigeria, Cameroon, Equatorial Guinea, Gabon, Congo, Democratic Republic of the Congo, Madagascar

1-northern; Uganda, Tanzania,Kenya
2-Serengeti National Park; Central African Republic
3-Serengeti; Gabon
4-Dar es Salaam (Tanzania); 14,787 feet

Week 17

1-Botswana, Zambia, and Malawi; Namibia, South Africa, Mozambique, and Madagascar
2-Cape of Good Hope; Namib Desert
3-farming and ranching; Madagascar
4-in the farthest south and farthest north; Congo River

1-South Africa; Namib Desert in Namibia and Kalahari Desert in Botswana
2-Drakensberg; Madagascar
3-no; Swaziland
4-Malawi; moderate with rainy summers

1-at the mouth of the OkavangoRiver in Botswana; mostly in Botswana
2-Mozambique; Mozambique Channel
3-Cape Town, South Africa;Zambezi River
4-Namib Desert; Madagascar

Week 18

1-herding, hunting and small-farming; Antarctica
2-Sahara Desert; east
3-Nile (Africa); near the equator
4-South America; Ural Mountains, Caucasus Mountains

1-Central America, Mexico, and the Caribbean countries; Welland Canal
2-in Brazil near Rio de Janeiro; Amazon rain forest
3-4; Vatican City, Monaco;
4-Mount Elbrus (Gora El'brus Russia); about 2.5 million square miles

1-4,145 miles (Nile); 29,028 (Mount Everest, Nepal/ China)
2-840,000 square miles (Greenland); 226.500 square miles (Madagascar)
3-Puerto Rico Trench (28,232feet deep); Caspian Sea (18,900 cubic miles of water)
4-Antarctica; Lake Baikil in Russia (5,315 feet deep)

ANSWERS

Week 19

1-Eurasia; Yes, the Sinai peninsula is a part of Egypt.
2-Red Sea, Arabian Sea; any four of the following: Bahrain, Qatar, United Arab Emirates, Oman, Yemen
3-Turkey, Iraq (Tigris); Turkey, Syria, and Iraq (Euphrates); Saudi Arabia
4-Turkey, Syria, Lebanon, Israel, and Egypt; Iran

1-Middle East; Gulf of Suez, Gulf of Aqaba
2-Persian Gulf, Gulf of Oman; Saudi Arabia, United Arab Emirates, Oman, Yemen
3-Lebanon; Euphrates
4-Turkey, Syria, Cyprus, Lebanon, Israel, Egypt, Iraq, Jordan; dry

1-Red Sea and Mediterranean Sea; Iran, Iraq, Kuwait, Saudi Arabia, Qatar, United Arab Emirates
2-land between Euphrates and Tigris Rivers; Egypt, Jordan, Lebanon, and Syria
3-Nicosia, Cyprus; Yemen
4-hills and low tablelands; semiarid

Week 20

1-Caspian Sea; Georgia
2-Caspian Sea; Uzbekistan
3-swimming; northern (or eastern)
4-west; mountainous

1-Aral Sea; Caspian Sea
2-Georgia, Azerbaijan; below
3-Tajikistan; mountainous
4-Uzbekistan; desert or dry

1-Azerbaijan, Iran,Turkmenistan, and Kazakhstan; Caucasus
2-Turkmenistan, Uzbekistan, and Kyrgyzstan; Kazakhstan
3-Pamirs; 92 feet below sea level
4-sparse grass; Aral Sea

Week 21

1-Kabul; Indus River
2-India; Sri Lanka
3-Indian; India
4-Himalayas; India, Bangladesh

1-India; Mount Everest (Nepal-China)
2-Pakistan; Indian Ocean
3-Nepal; Bangladesh
4-Himalayas; dry and mountainous

1-Islamabad, Pakistan; Arabian Sea
2-Western Ghats and Eastern Ghats; Deccan Plateau
3-Bhutan; Maldives
4-Bangladesh and India; Colombo, Sri Lanka

ANSWERS

Week 22

1-China; North Korea and South Korea
2-Russia; Ob River, Lena River
3-Mongolia; Beijing
4-Japan; Taiwan

1-Kara Sea, Laptev Sea, East Siberian Sea (Barents Sea is also north of the European part of Russia); Sea of Japan
2-at the east China Sea near Shanghai; North Korea and South Korea
3-southwest area of China; Japan
4-Taiwan Strait; Gobi Desert

1-Tibet; Moscow, Russia and Nakhodka, Russia
2-Japan; Gobi Desert
3-China; Sea of Okhotsk
4-Sea of Japan; Taipei, Taiwan

Week 23

1-north; Indonesia
2-Myanmar (Burma); Philippines
3-Thailand; Malaysia
4-Singapore; Myanmar (Burma)

1-Gulf of Thailand; Indonesia
2-Malaysia and Indonesia; Manila
3-Malaysia has land both east and west of this line (sneaky question!); Myanmar (Burma)
4-Jakarta; Philippines

1-Cambodia, Laos, Myanmar (Burma), Thailand, Vietnam, and western portion of Malaysia; Ho Chi Minh City
2-Java (in Indonesia); Philippines
3-Thailand, western Malaysia, and Singapore; Laos
4-Irrawaddy (Ayeyarwady) River in Myanmar (Burma); Hanoi

Week 24

1-Australia;Great Barrier Reef
2-Tropic of Capricorn; New Zealand
3-Sydney; sheep and wool
4-Tasmania; ranching

1-a country in Oceania; Coral Sea
2-southeastern; Indian Ocean
3-Aborigines; Cook Strait
4-Darling; moderate with humid or rainy summers

1-Australia, New Zealand, Papua New Guinea and islands of the South Pacific; Great Sandy Desert, Gibson Desert, Great Victoria Desert
2-in the southeast part; above
3-Great Barrier Reef; Outback
4-mild and rainy; Canberra, Australia

ANSWERS

Week 25

1-southern; north
2-yes; Canberra
3-Mount Kosciuszko; northern parts closest to the equator
4-Tropic of Cancer and Tropic of Capricorn; Great Sandy Desert, Great Victoria Desert

1-any one of the following: Melanesia, Micronesia, Polynesia; yes
2-Polynesia; Indonesia
3-Tropic of Capricorn; east
4-tropical climate; The Great Barrier Reef

1-Melanesia, Micronesia, Polynesia; Solomon Islands
2-New Guinea; Asia and Oceania
3-Mount Wilhelm, 14,793 feet, is located in Papua New Guinea; Hawaii (Mount Waialeale gets 460 inches a year!)
4-Wellington, New Zealand; Lake Eyre (52 feet below sea level) in south Australia

Week 26

1-129°F below zero in Antarctica at the Vostok Research Base); zero (Trick question! Antarctica has only visiting research scientists and tourists.)
2-Southern Ocean; ice and snow
3-South Pole; Ross Sea
4-Vinson Massif at 16,066 feet; Ronne Ice Shelf

1-South America; Eastern Hemisphere
2-frigid (polar ice cap); Transantarctic Mountains
3-No, it is north of the Antarctic Circle at 140°E; Amery Ice Shelf
4-Cape Norvegia; Ellsworth Mountains

1-Ross Sea and Ross Ice Shelf; Eastern Hemisphere
2-near 60°W toward South America; Polar Basin
3-Africa; near a point at 65°S and 140°E
4-Ross Sea, Amundsen Sea, Bellingshausen Sea, and Weddell Sea; Ross Ice Shelf

Week 27

1-Asia; Greenland
2-rain forest; desert
3-a land that has a government; Pacific Ocean
4-Mount Everest; Asia

1-one-third; Dead Sea
2-Bosporus Strait; rice
3-they have few natural enemies; wool
4-McMurdo Station; Drake Passage

1-Saudi Arabia; 20,230 feet (Mount McKinley, Alaska)
2-any three of the following: Iran, Afghanistan, Bhutan, Myanmar, Sri Lanka; 35,810 feet (Mariana Trench in thePacific Ocean)
3-more than 1,000 mph; 66,700 mph
4-Tokyo-Yokohama agglomeration (30 million people); China (over 1 billion)

Additional Resources

- Order Form -

Title	Price	Qty	Total
Trail Guide to World Geography	18.95		
Trail Guide to U.S. Geography	18.95		
Uncle Josh's Outline Map Book	19.95		
Uncle Josh's Outline Maps CD-ROM	26.95		
The Ultimate Geography and Timeline Guide	34.95		
US/World Mark-It Map (laminated)	9.95		
Eat Your Way Around the World	14.95		
Geography Through Art	19.95		
Geographical Terms Chart	4.75		
Beginner World Atlas	5.95		
Intermediate World Atlas	6.95		
Answer Atlas	12.95		
Around the World in Eighty Days	4.99		
World TG Student Notebook (All Three Levels CD-ROM)	34.95		
World TG Student Notebook (All Three Levels eBook)*	28.00		
Continents Map Set (large paper outline maps)	12.95		
Usborne Encyclopedia of World Geography	19.95		
Galloping the Globe	24.95		

Mail order with payment to:

Geography Matters
P.O. Box 92
Nancy, KY 42544

Subtotal ____________

S & H (12% of Subtotal $6 min) ____________

Tax: KY residents add 6% ____________

Total ____________

Ship To:

____
Name

____
Address

____
City/State/Zip

____
Phone

____
Email * please include your email address for eBook delivery

Payment Info:

Visa ❑ MasterCard ❑ Discover ❑ Check ❑

Payment Type (Check One)

— — —
____
Card Number

/
______________ ______________
Expiration Date Security Code #

____
Signature

All prices and availability are subject to change. Call or check online for current information.

800-426-4650 **orders@geomatters.com** **www.geomatters.com**

Seminar Presentations

THE JEWELS OF DISCOVERY LEARNING

Leaving No Stone Unturned

Cindy Wiggers is a national seminar speaker who presents captivating seminars on the jewels of discovery learning. Her information-packed workshops are loaded with clever ideas and activities designed to help teachers spur students on to discover knowledge for themselves. Cindy's presentations are practical, entertaining, and down-to-earth as she shares - from real life experiences - her heart to help children develop a delight for learning.

How to Teach Geography in a Way That Has Your Kids Asking for More

Are you encountering geography in your everyday life experiences? You can, and Cindy will show you how! Knowing geography's five themes will help you make this topic come alive in your daily learning activities. Discover creative ways to incorporate games, simple mapping assignments, and atlases and almanacs. Kids love these kinds of hands-on projects and will want to learn more about the geography of our world.

What Everybody Ought to Know about Timelines

Want to help your students remember historical events and their significance to history? Add a practical hands-on timeline to any history study and watch understanding, memory retention, and learning enjoyment increase immediately. Finally, a way to teach history with regard to chronology and cause and effect.

Notebooks: The Secret of Adding Spice to Any Curriculum

Did you know that George Washington, Benjamin Franklin, and Thomas Jefferson all kept personal learning notebooks? Students learn more from what they personally produce in their own notebooks. From the creative side to academics, your students can establish an excellent journal of their learning experiences while enjoying the benefits of self-education.

Teaching History and Geography with Literature

Learn how to make the most of your read-aloud time by extracting history and geography lessons and projects from nearly any historical novel. Students gain a more comprehensive understanding of history by learning the geography of the area in focus. Packed with practical ideas to help lead students to become more self directed learners.

7 Secrets to Help Kids Think for Themselves

Do you have trouble keeping your students interested in learning? Perhaps it's time you begin to think outside the box. Leonardo da Vinci approached knowledge from a unique perspective. Discover techniques used by da Vinci that, when implemented, will improve critical thinking and creativity, inspiring your students to want to gain knowledge. Included also in this workshop are ideas from the teachings of Ruth Beechick to develop thinking skills which help create a delight for learning and improves your children's ability to excel in any area they choose to pursue.

To schedule any of these presentations at your convention or group function, contact Cindy at cindy@geomatters.com or 606-636-4678.